WHAT IS TIME?

What is Time?

G. J. WHITROW

With a new introduction by J. T. Fraser,
and a new bibliographic essay by J. T. Fraser and M. P. Soulsby

OXFORD
UNIVERSITY PRESS

OXFORD

UNIVERSITY PRESS

Great Clarendon Street, Oxford OX2 6DP

Oxford University Press is a department of the University of Oxford.
It furthers the University's objective of excellence in research, scholarship,
and education by publishing worldwide in

Oxford New York

Auckland Bangkok Buenos Aires Cape Town Chennai
Dar es Salaam Delhi Hong Kong Istanbul Karachi Kolkata
Kuala Lumpur Madrid Melbourne Mexico City Mumbai Nairobi
São Paulo Shanghai Taipei Tokyo Toronto

Oxford is a registered trade mark of Oxford University Press
in the UK and in certain other countries

Published in the United States
by Oxford University Press Inc., New York

First published 1972 by Thames & Hudson Ltd., London
First published as an Oxford University Press paperback,
with a new introduction and bibliographic essay 2003

British Library Cataloguing in Publication Data

Data available

Library of Congress Cataloging in Publication Data

Data available

ISBN-13: 978-0-19-860781-6
ISBN-10: 0-19-860781-4

2

Typeset in Photina MT by RefineCatch Limited, Bungay, Suffolk
Printed in Great Britain by Clays Ltd, St Ives plc

To Magda

CONTENTS

INTRODUCTION

'On ye sholders of Giants'

J. T. Fraser

On 5 February 1676, in the course of his correspondence with Robert Hooke concerning the nature of light, Newton penned a letter to his fellow physicist.

> [Y]ou defer too much to my ability for searching in this subject. What Des-Cartes did was a good step. You have added much several ways, & especially in taking ye colours of thin plates into philosophical consideration. If I have seen further it is by standing on ye sholders of Giants.[1]

By Newton's time the aphorism of the giants' shoulders was over five centuries old. It seems to have originated in the writings of Bernard of Chartres, an early twelfth-century humanist philosopher. During the history of the West, from age to age, the aphorism kept on reflecting and refracting in intellectual discourse, taking slightly different forms. All along, it stood for the recognition of the cumulative nature of human knowledge. A diligent student of early mankind might be able to trace its history to the first child who wanted to be lifted to his father's shoulders because from there, and only from there, could he see the pretty bird in the faraway bush.[2]

---

1. Turnbull, H. W. (ed.). *The Correspondence of Isaac Newton*, Vol. 1. Cambridge University Press, 1959, p. 416. Courtesy of the Burndy Library, Dibner Institute for the History of Science and Technology, Cambridge, MA.
2. For the amusing and amazing story of this aphorism – not including the child on his father's shoulders – see Robert K. Merton's *On the Shoulders of Giants*, San Diego: Harcourt Brace Jovanovich, 1965.

The temporal dimension of the world, as perceived by our most distant human ancestors, had to resemble the time perception of their immediate, primate forebears. I imagine that these hairy humans were able to prepare for the cyclic and, therefore, predictable aspects of their near future but, for the meeting of contingencies, they had available only such skills as natural selection had distilled in their biological endowment. I also imagine that as the store of their responses to contingencies widened, and with it the temporal horizons of their reality broadened, they acquired a certainty about the inevitable end to their selves. They became concerned with survival after death and with that concern, they became fully human.

Then followed an immense journey, an opening of the spatial and temporal horizons of their cosmos. Along that journey, a mere twenty-four centuries ago, a form of disciplined speculation, called philosophy, was born. Centuries later, many of the early philosophers came to be seen as practitioners of natural philosophy because their reasoning was rooted in their observation and testing of nature.

'The history of natural philosophy', wrote G. J. Whitrow in his magistral *The Natural Philosophy of Time*, 'is characterized by the interplay of two opposing points of view which may be conveniently associated with the names of Archimedes and of Aristotle, those intellectual giants of antiquity whose writings were of decisive importance for the late medieval and Renaissance founders of modern science.' Archimedes, continued Whitrow, is a prototype of those thinkers who believe that the passage of time is not an intrinsic, ultimate basis of things, whereas Aristotle is the forerunner of those who regard time as fundamental to the universe.[3]

As did Bernard's aphorism about knowledge, ideas about the nature of time also reflected and refracted throughout history. They were born and died, they were formulated, reformulated, and modulated by each culture, each religion, each philosophy, each science,

---

3. Whitrow, G. J. *The Natural Philosophy of Time*, 2nd edn. Oxford: Clarendon Press, 1980, p. 1.

each art, each person. In their many forms they served as models of the human experience of passage in the presence of, and in contrast to, the fantasy of deathless life.

The human experience of time is all-pervasive, intimate, and immediate. Life, death, and time combine in an intricate and intriguing manner that is difficult to clarify, yet it is recognized in all great philosophies and religions. Time is a constituent of all forms of human knowledge, all modes of expression, and is connected with the functions of the mind. It is also a fundamental feature of the universe. It follows that no single faculty of learning, in itself, is capable of accounting for the nature of time. The consequent intensity of concern and argumentation is understandable.

Along their strange and eventful history, using their intelligence, diligence, and ruthlessness, and employing their knowledge of time as a weapon, the creatures that buried their dead and became concerned with survival after death made the whole earth their home. By the twenty-first century they made that home into a global laboratory in which they experiment with time-compactness in the lives of persons, nations, and cultures. Time-compactness communicates itself to each and every facet of personal and collective life. It enters the formulation of principles of conduct and value judgment and sets the tone in which we say and do things. It insinuates itself into all inquiries, secular and spiritual, and amplifies cultural, socioeconomic, ecological, and ideological crises.

In the face of these crises, all prior teachings about the position of man in the universe, and all prior ideas about the nature of time, came to be questioned. The many ways in which time relates to man, animal, and matter have been reexamined. Scores of time-related conferences have been held, hosts of books and papers published, with themes spread across the spectrum of abstract and applied knowledge. The fertile heterogeneity of ideas that emerged and the intensity of debates that took place suggest the natural philosophy of time as a suitable framework for surveying the immensity of contemporary knowledge.

In its mature form, a natural philosophy of time should offer a coherent, intellectual structure that respects and accommodates the many and different modes of reasoning, testing for truth, and formulating lawfulness, that the disciplines that contribute to it have found necessary to employ. The natural philosophy of time is in the process of doing just that: it is developing a vocabulary of concepts and a body of principles that allow it to deal with time felt and time understood, as well as with the many issues of passage and permanence in humans, animals, and matter.

Whitrow's remarkable achievement is that he had the learning, combined with intellectual courage, to handle within one work – *The Natural Philosophy of Time* – all those details of the many departments of humanities and the many faculties of the sciences which, in his judgment, were necessary for an interdisciplinary study of time.

As does *The Natural Philosophy of Time*, so the present volume, *What is Time?*, abides by the Heraclitean dictum that 'men who love wisdom must have knowledge of very many things'.[4] It recognizes that our understanding of the nature of time must remain partly obscured until we learn to benefit from insights that come from all our rational, introspective, and experimental knowledge.

That kind of an inquiry is no mean task because, in our epoch, specialists in different areas of knowledge occupy separate islands and because the inhabitants of those islands are, so to say, not on speaking terms. Beyond problems of their different jargons and criteria of values, their separation and alienation is aggravated by the fact that inquiries into the nature of time are never emotionally neutral. Debates about time tend to become parochial and narrowly dogmatic. The reason is that thinking about time mobilizes the conflicts between our certainty of passage and the fantasy of eternal life. In that mobilized turmoil, beyond the field and wave equations, beyond evolution by natural selection, beyond psychology and soci-

---

4. This is Heracleitus' 'On the Universe', Fragment XLIX. In *Hippocrates – Heracleitus on the Universe*, Vol. 4, trans. W. H. Jones. Loeb Classical Library. Cambridge, MA: Harvard University Press, 1992, p. 487.

ology, and beyond the arts, the letters, and religions, there always remains a rational mystery about passage, rooted in the incompletability of nature.

In the company of Whitrow's other time-related writings, and of the extensive body of works by others, of which the bibliography appended to this book is only a thin excuse of a sampling, what are the merits of *What is Time?*

This modest volume is remarkable in the clarity of its exposition. It is inspiring in the meticulous reliability of its material. Also, unlike many popular works about time, it is without hard sell: it does not propose either to seat or to unseat the Almighty. Few are the authors who, like Whitrow, could combine substantial humanistic scholarship with a solid command of clear scientific thought and, by appealing to them, could deal with the many ways that humans employ in their negotiations among themselves and with their environment.

The way to pursue the ambitious task envisaged by G. J. Whitrow is to support an intellectual climate where creativity common to all knowledge is permitted to flourish, and where aspects of reality, separately understood, are permitted to produce an integrated body of understanding – by interacting through the idea of time. This is a difficult task. But difficulties never stopped scholars and scientists from pursuing their goals of unified, structured visions. They have known that in the countryside of the intellect and culture, as in the countryside of New Hampshire – in the words of the poet Robert Frost – although 'good fences make good neighbors',[5] still,

> Something there is that doesn't love a wall,
> That sends the frozen ground swell under it,
> And spills the upper boulders in the sun.

---

5. 'Mending wall', in *The Poetry of Robert Frost*. New York: Holt, Rinehart & Winston, 1969, p. 33.

PREFACE

Late in 1969 I gave four talks in the Third Programme of the BBC on the nature of time. They were published in the *Listener* in January 1970. The present book is a considerably expanded version of these talks.

I wish to take this opportunity of thanking Christopher Sykes who originally commissioned the talks and edited the first draft of the scripts, my producer Daniel Snowman, and also Thomas Neurath who kindly invited me to write the present book. As always, I owe a great debt to my wife Magda Whitrow for her critical comments and for compiling the index. I should also like to thank Jane Galletti for her elegant and careful typing of the manuscript.

G. J. W.
September 1971

1

The origin of our idea of time

THE STORY IS TOLD of the Russian poet Samuel Marshak that when he was first in London, before 1914, and did not know English well he went up to a man in the street and asked, 'Please, what is time?' The man looked very surprised and replied, 'But that's a philosophical question. Why ask me?' Many centuries ago a famous Church Father was troubled by the same question and confessed that if no one asked him he knew, but if he tried to explain it to someone then he had to admit that he did not know. Although there are many important ideas that most of us agree we do not understand, only time has this peculiar quality which makes us feel intuitively that we understand it perfectly so long as we are not asked to explain what we mean by it.

The object of this book is to discuss the nature of time from various points of view. The first question to consider is the origin of the idea that time is a kind of linear progression measured by the clock and calendar. In modern civilization this conception of time so dominates our lives that it seems to be an inescapable necessity of thought. But this is far from true. Not only do primitive races have only extremely vague ideas about clocks and calendars but most civilizations, prior to our own of the last two or three hundred years, have tended to regard time as essentially cyclic in nature. In the light of history, our conception of time is as exceptional as our rejection of magic.

Although our conception of time is one of the peculiar characteristics of the modern world, the importance that we attach to it is not entirely without cultural precedent. Nor indeed is our present Gregorian calendar – named after Pope Gregory XIII who introduced it in March 1582 – the most precise that any civilization has used. Our calendar, for all its sophistication, is not quite as accurate as that devised by the Maya priests of Central America more than a thousand years ago. The Gregorian year is slightly too long, the error amounting to three days in ten thousand years. The length of the year according to the Maya astronomers was too short, but the defect amounted to only *two* days in ten thousand years.

Of all people known to us, the Mayas were the most obsessed with the idea of time. Whereas in European antiquity the days of the week were regarded as being under the influence of the principal heavenly bodies – Saturn-day, Sun-day, Moon-day and so on – for the Mayas each day was itself divine. Every monument and every altar was erected to mark the passage of time, and none was for the glorification of rulers or conquerors. The Mayas pictured the divisions of time as burdens carried by a hierarchy of divine bearers who personified the respective numbers by which the different periods – days, months, years, decades and centuries – were distinguished.

Despite their constant preoccupation with temporal phenomena, and the amazing accuracy of their calendar, the Mayas never attained the idea of time as the journey of one bearer with his load. Their conception of time was magical and polytheistic. Although the road along which the divine bearers marched in relays had neither beginning nor end, events moved in a circle represented by the recurring spells of duty for each god in the succession of bearers. Days, months, years and so on were all members of relay teams marching through eternity. Each god's burden came to signify the particular omen of the interval of time in question. One year the burden might be drought, another a good harvest. By calculating which gods would be marching together on a given day, the priests could determine the combined influence of all marchers and thus forecast the fate of mankind.

The hierarchy of cycles for each division of time led the Mayas to devote more attention to the past than to the future. History was expected to repeat itself in cycles of 260 years, and significant events would tend to follow the preordained general pattern. For example, the Christian religion introduced by the Spaniards was identified with the worship of an alien cult that had been imposed on the Mayas several centuries earlier. Past, present and future events blended in the Maya world-view because they all resulted from the same divine burden of the 260-year cycle.

To us this confusion of past, present and future by a highly intelligent people, who made a sustained and sophisticated mathematical study of astronomical time, seems very strange indeed, particularly as we now have abundant evidence that our sense of these temporal distinctions is one of the most important mental faculties distinguishing men from all other living creatures. It seems that all animals except man live in a continual present. Such examples as may be cited to the contrary do not survive critical investigation. Dogs frequently display powers of memory when they give vent to the wildest joy on seeing their masters after long separation; but this does not necessarily indicate any image of the past as such. Similarly, there is no firm evidence that any animals have a sense of the future. Carefully analysed experiments have shown that, even in the case of the most intelligent animals such as chimpanzees, any actions they take that might be thought to indicate some such sense are in fact purely instinctive.

In man's case, awareness of the distinctions between past, present and future must have been the result of conscious reflection on the human situation. The mental and emotional tension resulting from man's discovery that every living creature is born and dies, including himself, must have led him intuitively to seek some escape from the relentless flux of time. There is evidence that even Neanderthal Man, the precursor of *Homo sapiens*, buried his dead and may even have provided for what he imagined to be their future needs. As for our own species, the oldest evidence, going back to roughly 35,000 BC, reveals that ritual burial was already an established custom. The dead were not only equipped with weapons, tools and ornaments but even with food, which often must have been in short supply for the living. Burials have even been found in which the body was covered with some red pigment, no doubt with the magical intent of endowing the dead person with the colour of life-giving blood in the hope of averting his physical extinction. Usually the body was buried in a crouched posture, which may have been inspired by the idea that the dead were being placed in the womb of Mother Earth for future rebirth. Whether this explanation of our remote ancestors' burial customs is valid is, of course, an open question, but it may be an indication of the origins of the cyclic conception of life.

An enormous effort must have been required for man to overcome his natural tendency to live like the animals in a continual present. From our knowledge of surviving primitive races we have ample evidence of this. For example, although the children of Australian aborigines are of similar mental capacity to white children, they have great difficulty in telling the time by the clock. They can read off the position of the hands on the clock as a memory exercise, but they are quite unable to relate it to the time of day. There is a gap, which they find difficult to cross, between their conception of time and that of modern industrial civilization. It was surely significant that Rousseau, who extolled the noble savage, detested time and threw away his watch.

A vital factor in man's primitive intuition of time was his sense of rhythm. A highly developed sense of rhythm enabled a tribe to function with precision as a single unit both in war and in hunting. Time is also experienced by man in the periodicity of his own life as well as in the periodicity of the natural world. The principal transitions from one phase of a man's life to another were thought of as crises and as a result the community to which he belonged assisted him with the appropriate rituals. Similarly, the principal transitions in nature were also regarded as occurring suddenly and dramatically. With the change from a pastoral and nomadic to an agricultural and more highly organized form of existence, the importance of the cyclic phenomena of nature must have been enormously enhanced. Nature was seen as a process of strife between divine cosmic powers and demoniacal chaotic powers in which man was not just a spectator but was obliged to play an active part in helping to bring about the required phenomena by acting in full unison with nature. This meant acting out a given set of rituals. Thus, for over two thousand years, right down to Hellenistic times, the Babylonians celebrated a New Year's festival of several days – which occurred roughly at the time of the spring equinox – during which the story of creation was re-enacted and even a mock-battle was fought in which the king impersonated the victorious god. In Egypt, where everything depended on the Nile, the coronation of a new Pharaoh was made to coincide either with the rising of the river in early summer or with the recession of the waters in autumn when the fertilized fields were ready to be sown.

The creation epic liturgically recited by the chief priest of ancient Babylon was not really regarded as a record of *the past*. Instead, it served the theologico-politico purpose of securing the god Marduk's supremacy in *the present*. For Marduk, who was not the most ancient of the gods, was the particular deity associated with Babylon and his lordship over the other gods was meant to justify the political supremacy acquired by that city.

Similarly, the Egyptians were no more historically minded than the ancient Mesopotamians. Nevertheless, in one respect they made an outstanding contribution to the science of time. For they devised what Otto Neugebauer has described as 'the only intelligent calendar which ever existed in human history'. Their year consisted of twelve months, each of thirty days, with five additional days at the end of the year. It is thought to have originated on purely practical grounds from continual observation and averaging of the time intervals between successive arrivals of the Nile flood at Cairo.

Despite the Egyptian example, calendars originally tended to be primarily associated with religion, because it was important for feasts and sacrifices to be celebrated on fixed dates. Why should God mind exactly when Easter is celebrated? As we have already seen in Babylonia, the king-priest was the incarnation of the invisible god in the sky and the rituals he performed were the repetition of divine actions and had to correspond exactly in time as well as in character with the rituals on high. To this primitive idea we trace back the importance of celebrating Easter at the correct date, for this was the crucial time of combat between God (or Christ) and the Devil, and God required the support of his worshippers to defeat the Devil.

The importance of celestial influences on the ideas of time and the calendar are due in origin to the Chaldeans, or late Babylonians. Their astrology was based on the fundamental assumption that all events on Earth are influenced by the stars. In particular, we can trace back to them, by way of the Hebrews, the origin of our present seven-day week associated with the Sun, Moon and the five planets that they discovered. (The separation of these planets from the so-called 'fixed stars' was one of their greatest achievements.)

The planetary week presents a strange combination of ideas from different cultures. From Babylon came the doctrine of the influence of the stars on man's destiny, from the Alexandrian Greeks came the mathematical astronomy that placed the planets in a certain order of distance from the Earth, and then on these foundations the late Hellenistic astrologers, who were familiar with the ancient cult of the magical number seven, constructed a purely pagan week. By the end of the third century AD, the Christians, who had previously adhered to the Jewish seven-day week in which the days simply had numbers and not names, began to be influenced by the astrological beliefs of converts from paganism and changed over to the planetary week. The stars were no longer regarded as deities but as demons capable of affecting the fate of man. At the same time, the oriental worship of the Sun-god Mithra was extremely influential in the Roman world. This led to the substitution by pagans of the *dies Solis* (the Sun-day) for the *dies Saturnis* (the Saturn-day) as the first day of the week. This change appealed to the Christians, who had long observed Sunday – the Lord's Day (*dies Dominica*), on which Christ rose from the dead – as the first day of the week, in place of the Jewish Sabbath. Incidentally, the earliest Sunday law appears to have been embodied in an edict of the Emperor Constantine in AD 321 enacting that magistrates, citizens and artisans were to rest from their labours 'on the venerable day of the Sun'. In the same century, the fourth, Christmas was assigned to 25 December, because on that date each year was celebrated the birth of the Sun to a new life after the winter solstice. Easter, being a lunar feast in origin (occurring at or just after full moon), retained a variable date.

The influence of Christianity on our modern concept of time is not restricted to calendrical details. It was far more fundamental than that. Its central doctrine of the Crucifixion was regarded as a unique event in time not subject to repetition, and so implied that time must be linear and not cyclic. Belief in the cyclic pattern of time was a common feature of many ancient cultures and in particular characterized Greek cosmological ideas, especially in Hellenistic times. It found its apotheosis in the idea of the 'Great Year' so vividly described by Nemesius, the fourth-century bishop of Emesa:

> The Stoics say that when the planets return, at certain fixed periods of time, to the same relative positions which they had at the beginning, when the kosmos was first constituted, this produces the conflagration and destruction of everything which exists. Then again the kosmos is restored anew in a precisely similar arrangement as before. The stars again move in their orbits, each performing its revolution in the former period, without variation.

According to Nemesius, the Stoics even believed that

> ... Socrates and Plato and each individual man will live again, with the same friends and fellow-citizens. They will go through the same experiences and the same activities. Every city and village and field will be restored, just as it was. And this restoration of the universe takes place not once, but over and over again – indeed to all eternity without end. Those of the gods who are not subject to destruction, having observed the course of one period, know from this everything which is going to happen in all subsequent periods. For there will never be any new thing other than that which has been before down to the minutest detail.

Before the rise of Christianity, with the exception of a few isolated writers like Seneca, only the Hebrews and the Zoroastrian Iranians appear to have thought of history as progressive rather than cyclic. The eschatology of the early Hebrew prophets was no doubt greatly influenced by the fate of Israel as a state after it was conquered by the Babylonians, the future alone holding the promise of well-being for the community of faithful Israelites. In short, the essential aim of the Jewish God in history was the salvation of Israel. The key book in the Old Testament for our present purpose is, however, one written under the later stress of danger from the Seleucids, just before the Maccabean rising: the *Book of Daniel*, where history was presented, under the guise of prophecy, as a unified process conforming to a divine plan of teleological significance.

The religion that was in strongest competition with Christianity in the early centuries of the Christian era was also greatly concerned with the significance of time. For Mithraism – many features of which were ultimately absorbed by Christianity – was derived from the heretical form of Zoroastrianism known as Zurvan. To members of this sect time (*Zurvan*) was the source of all things and was the father of the twin spirits of good and evil, Ohrmazd and Ahriman, who alone figured as primary in the original dualism of early Zoroastrianism. A distinction was made between *Zurvan akarana* (infinite time) and 'Time of the Long Dominion' (finite time). The latter, lasting twelve thousand years (the number twelve being associated with the twelve signs of the Zodiac), is the period of the struggle between the spirits of good and evil. In fact the whole *raison d'être* of finite time appears to have been to bring about that conflict of good and evil which leads to the ultimate triumph of the good. Although finite time was conceived as moving in a circle, this circular movement was not regarded as eternal. The Iranian theory of time, therefore, had little or no affinity with the *Aion* speculations (eternity perpetually repeating itself, see p. 10) of the Hellenistic world or the ever-recurring *Kalpas* (successive cycles) of the Hindus. Instead, at a given moment finite time comes into existence out of infinite time, moves in a circle until it returns to its beginning, and then merges into infinite time − that is, timelessness. The process is never renewed.

The Christian view of time is more thoroughly linear and freer from cyclic conceptions than even the Zoroastrian and Hebrew. For emphasis on the non-repeatability of events was the very essence of Christianity. The contrast with the Hebrew view is clearly brought out in the *Epistle to the Hebrews*, Chapter 9, verses 25 and 26: 'Nor yet that he should suffer himself often, as the high priest entereth into the holy place every year with the blood of others; For then must he often have suffered since the foundation of the world; but now once in the end of the world hath he appeared to put away sin by the sacrifice of himself.' It is true that the idea of denominating the years by a single era count had been adopted by the Greeks in the third century BC, when the historian Timaeos introduced the device of dating by the Olympiads from 776 BC. But in the Christian era the AD sequence was not adopted until the year 525. As for the BC sequence, extending backwards from the birth of Christ, this was not introduced until the seventeenth century. This was because in medieval Europe, as in antiquity, time was not conceived as a continuous variable but was split up into several seasons, divisions of the Zodiac, and so on, each exerting its specific influence. In other words, magical time had not yet been succeeded by scientific time.

Throughout the whole medieval period the cyclic and linear concepts of time were in conflict. Scientists and scholars, influenced by astronomy and astrology, tended to emphasize the cyclic concept. The linear concept was fostered by the mercantile class and the rise of a money economy. For as long as power was concentrated in the ownership of land, time was felt to be plentiful and was associated with the unchanging cycle of the soil. But with the circulation of money the emphasis was on mobility. The tempo of life was increased, and time was now regarded as something valuable that was felt to be slipping away continually: after the fourteenth century public clocks in Italian cities struck all twenty-four hours of the day. Men were beginning to believe that 'time is money' and that one must try to use it economically.

One of the most important cultural consequences of man's changing attitude to time in the late Middle Ages and High Renaissance was its effect on the visual arts, causing painting *a secco* to replace *a fresco*, or true fresco. For the very long apprenticeship that pupils had to serve before they became proficient in fresco painting could not be maintained when social changes and pressures stimulated the desire for speed. Whereas, formerly, an artisan could linger over the execution of his work, a painter who had risen in the social scale, and had thereby acquired a new prestige, had to work fast in order to handle all the commissions he received. Even Michelangelo's example was in vain. Originally, it had been planned that the *Last Judgement* in the Sistine Chapel should be painted *a secco* in oil, but Michelangelo objected that oil painting is only 'fit for women and slovenly people' and so he carried out the work *a fresco*. But this proved to be against the current of the age and the glorious art of the true fresco died out, its practice being incompatible with the new social attitude to time.

In a fascinating essay[1] on the iconology of 'Father Time', the celebrated art-historian Erwin Panofsky has drawn attention to the contrast between the symbolic representations of time in classical art – either as fleeting opportunity (*Kairos*) or as creative eternity (*Aion*) – with the typical Renaissance image of time as the destroyer, equipped with hour-glass, scythe or sickle. No period, he argues, has been so obsessed with the horror and the sublimity of time as the Baroque, 'the period in which man found himself confronted with the infinite as a quality of the universe instead of as a prerogative of God'. This obsession with the destructive aspect of time can be seen in Shakespeare, notably in his sonnets and in the *Rape of Lucrece*, as in stanza 133:

> Mis-shapen Time, copesmate of ugly Night,
> Swift subtle post, carrier of grisly care,
> Eater of youth, false slave to false delight,
> Base watch of woes, sin's pack-horse, virtue's snare,

---

1. E. Panofsky, *Studies in Iconology, Humanistic Themes in the Art of the Renaissance* (the Mary Flexner lectures, 1937) Oxford, 1939.

> Thou nursest all and murder'st all that are:
> O, hear me then, injurious, shifting Time!
> Be guilty of my death, since of my crime.

Intellectual obsession with time dominated the thought of Shakespeare's contemporary Edmund Spenser. A fantastic example of this influence is his marriage ode *Epithalamion*. Some ten years ago, in a masterpiece of literary detection,[2] an American professor, Kent Hieatt, showed that this well-known poem is a tissue of time-symbolism in which the movement of the Sun throughout the day and the year are indicated in detail. Not only do the twenty-four stanzas represent the number of hours in the day but the long lines are symbolic equivalents of the duration of time and the short ones of the division of time. He pointed out that the total number of long lines is 365, the number of days in the year. Other line-totals imitate the apparent motion of the Sun throughout the year relative to the fixed stars. The poem was written to celebrate Spenser's own wedding, which occurred in southern Ireland at mid-summer. At that time in that latitude the number of hours of daylight is about sixteen and three-quarters. This is actually indicated in the poem by a change in refrain in the seventeenth stanza where night falls.

In Spenser we find the cyclic and linear concepts of time vying with each other. On the one hand, his idea of time was dominated by the figure of the everlasting wheel of change – a universal cyclical harmony by which all things return to themselves and temporal beings achieve permanence through constant regeneration. But, side by side with this age-old idea, Spenser's religious conviction led him to conclude that mutability will not last forever but will give way to a truly eternal perfection when time itself shall cease:

> But time shall come that all shall changed bee,
> And from thenceforth, none no more change shall see.

---

2. A. Kent Hieatt, *Short Time's Endless Monument: the Symbolism of the Numbers in Edmund Spenser's 'Epithalamion'* (New York, 1960).

For all his learning, Spenser was essentially a backward-looking figure and should be regarded as one of the great exponents of traditional cyclic ideas concerning time. The great leaders of the scientific revolution of the seventeenth century were also much concerned with temporal and horological metaphors, but with them a different outlook was beginning to prevail. Early in the century Kepler specifically rejected the old quasi-animistic magical conception of the universe and asserted that it was similar to a clock, and later the same analogy was drawn by Robert Boyle and others. Indeed, the invention of the mechanical clock played a central role in the formulation of a mechanistic conception of nature that dominated natural philosophy from Descartes to Kelvin. An even more far-reaching influence has been claimed for the mechanical clock by Lewis Mumford, who has argued that it 'dissociated time from human events and helped create the belief in an independent world of science'.

As regards the influence of the mechanical clock on the concept of time itself, an important feature distinguishes this kind of time-keeper from its predecessors. The oldest modes of time-reckoning were essentially discontinuous. For, instead of depending on a continuous succession of temporal units, they merely involved the repetition of a concrete phenomenon occurring within a unit – as, for example, where Homer, in the Twenty-first Book of the *Iliad*, makes one of Priam's sons say to Achilles, 'This is the twelfth dawn since I came to Ilion.' Even the sundials, sand-reckoners and water-clocks of antiquity were more or less irregular in their operation, and it was not until a successful pendulum clock was invented by the Dutch scientist Christian Huygens in the middle of the seventeenth century that man was at last provided with an accurate time-keeper that could tick away continually for years on end. This greatly influenced the modern concept of the homogeneity and continuity of time.

Meanwhile, during the course of the same century, the cyclical ideology began to give way to the new concept of linear progress professed as early as 1602 by Francis Bacon in an early work that bore the significant title (in Latin) *The Masculine Birth of*

Time. Nevertheless, even Newton adhered to the cyclical view and was convinced that the world was coming to an end. He believed that the comet of 1680 had just missed hitting the Earth, and in his commentaries on *Revelations* and the *Book of Daniel*, unpublished in his lifetime, he indicated that the end of the world could not be long delayed. A particularly striking example of his cyclical philosophy occurs in a letter that he wrote to Henry Oldenburg, Secretary of the Royal Society, in December 1675. 'For nature', he wrote, 'is a perpetual circulatory worker, generating fluids out of solids, fixed things out of volatile, and volatile out of fixed, subtle out of gross, and gross out of subtle, some things to ascend and make the upper terrestrial juices, Rivers and the Atmosphere; and by consequence others to descend for a Requitall to the former. And as the Earth, so perhaps may the Sun imbibe this Spirit copiously to conserve his shineing, and keep the Planets from recedeing further from him.'

The concept of linear advancement, however, was espoused by Leibniz, Barrow and Locke, among others. In the eighteenth century the new forward-looking view of time inspired the philosophers of the Enlightenment who abandoned the Biblical chronology that automatically excluded the possibility of slow processes of transformation over immense periods of time.

The idea of cosmic evolution which has dominated modern thought can be traced back to Descartes. Unlike Newton, who used his theory of gravitation to explain how the orbital motions of the planets and satellites can be maintained but not how they may have originated, Descartes assumed that originally the world was filled with matter distributed as uniformly as possible and he sketched out qualitatively a theory of successive formation of the Sun and planets.

Descartes's idea of the universe evolving by natural processes of separation and combination was the source of a succession of theories of cosmic evolution by Swedenborg, Buffon and others, but the first to apply Newtonian ideas to the problems of cosmogony was Immanuel Kant in his *Universal Natural History and Theory of the*

Figure 1: Kant's theory of how the solar system originated. A giant cloud of gas, contracting under its own gravitation, begins to rotate and sheds matter from its centre to form the planets by further gravitational contraction. In his view, the universe becomes less homogeneous with the passage of time.

Heavens, published in 1755. Kant began with the idea that in the beginning all matter was in a gaseous state and was spread more or less uniformly throughout the universe. Consequently he assumed, as many have done since, that we live in an evolutionary, or developing, universe in the sense that the past was essentially simpler than the present.

Although the idea of evolution was in the air, one of the obstacles that it had to contend with was the widespread inherited conviction that the range of past time was severely limited. Archbishop Ussher about 1650 calculated that God created the world on Sunday 23 October, 4004 BC. The calculation was somewhat less precise than the result would seem to indicate. The year 4004 BC was arrived at by taking Luther's estimate of 4000 BC, obtained by rounding off various arithmetical calculations of Biblical chronology, and then correcting it by four years to allow for Kepler's dating of the birth of Christ in 4 BC, because of a four-year error that he detected in the date of the Crucifixion when he compared New Testament dating with that of solar eclipses.

During the course of the eighteenth century scientists and others began to discard the Bible-based chronology of nature. In 1721, Montesquieu wrote in his *Lettres Persanes*, 'Is it possible for those who understand nature and have a reasonable idea of God to believe that matter and created things are only 6,000 years old?' In mid-century,

Diderot thought in 'millions of years' and Kant suggested that the universe may be hundreds of millions of years old.

In 1788, the geologist James Hutton in his *Theory of the Earth* rejected the sudden catastrophic agencies that had previously been invoked to explain the stratification of rocks, the deposition of oceans, etc. He realized that the true scientific approach is not to invoke such *ad hoc* hypotheses but to test whether or not the same agents as are operating now could have operated all through the past. In his view, the world has evolved and is still evolving (in one place he actually likened it to an organism). He concluded that vast periods of time were required for the Earth to have reached its present state and from his study of sedimentary and igneous rocks he concluded that 'We find no vestige of a beginning – no prospect of an end.'

Although during the eighteenth century increasing importance came to be attached to the idea of historical progress, cyclical conceptions of history tended to persist. In his famous *Scienza Nuova* (1725) Vico based his theory of history on his 'law of cycles'. The linear view of time as continual progression without cyclical repetition finally prevailed through the influence of the nineteenth-century biological evolutionists. Today, in a world dominated by Western science – in which even our everyday life is regulated chronometrically so that we tend to eat and sleep, not when we feel hungry or tired, but when prompted by the clock – our conception of time is the dominant feature of our world-view. It is therefore not surprising that it is generally assumed to be intuitively obvious, but as we have seen from history this is far from being the case.

2

Time and ourselves

THE INCREASING IMPORTANCE ATTACHED TO the idea of time following the scientific revolution of the seventeenth century led philosophers to investigate the nature of the concept and its role in our personal awareness of phenomena. A central figure in the study of these questions was the philosopher Immanuel Kant, who came to the conclusion that time is one of the forms of our 'intuition'. That is to say, it does not characterize external objects but only the subjective mind that is conscious of them. Consequently, Kant believed that the idea of scientific linear time was an automatic consequence of the fact that we are rational creatures. But we have already seen that this idea was explicitly formulated only in the seventeenth century, and previous civilizations had different concepts of time. Kant's conclusion therefore cannot be historically correct.

Towards the end of last century it was shown that Kant's theory of time was equally unsatisfactory from a psychological standpoint. In a famous essay[1] on the development of our concept of time, the French psychologist Guyau argued that time should not be regarded as a prior condition, but as a consequence, of our experience of the world, the result of a long evolution. Guyau argued that in the period of primitive mental confusion the succession of ideas in the mind did not automatically give rise to the idea of their succession. He rejected the naïve assumption of Herbert Spencer that the idea of

1. M. Guyau, *La Genèse de l'idée du temps* (Paris, 1890).

time was derived from a primitive awareness of temporal sequence and maintained that in his primitive state man had no clear conception of either simultaneity or succession. He suggested that the idea of time arose when man became conscious of his reactions towards pleasure and pain and of the succession of muscular sensations associated with these reactions. Guyau argued that whereas man's spatial conceptions may have originated when he became fully conscious of, and reflected upon, his movements, temporal concepts can be traced back to the feelings of effort and fatigue associated with these movements.

Guyau assumed that the human mind has the power, apparently not possessed by animals, to construct the idea of time from our recognition, or awareness, of certain features characterizing the data of our experience. Although Kant threw no light on the origin of this power, he recognized that it was a peculiarity of the human mind. In recent years it has become clear that all man's mental abilities are potential capacities which he can only realize in practice by learning how to use them. For, whereas animals inherit various particular patterns of sensory awareness – known as 'releasers' because they automatically initiate specific types of action – man has to learn to construct all his patterns of awareness from his own experience. According to Kant, our ideas of space and time function as if they were releasers. Instead, they should be regarded as mental apparatus that we have to learn to construct for ourselves.

Over many years much effort has been devoted to the elucidation of the physiological and psychological bases of our awareness of time. Traditionally, we regard our bodies as endowed with the three physical senses of sight, hearing and touch and the two chemical senses of taste and smell, but are we not also endowed with some sense of direct temporal awareness? On this question the most diverse views have been held. For example, the Austrian physicist and philosopher of science Ernst Mach argued in 1865 that we have a specific sensation of time associated with what he called the 'work of attention' – that is, with the effort that we make in concentrating our attention on one thing after another. A quarter of a century later Guyau maintained that this sensation, if it exists, is vague, irregular and exceedingly subject to error, but a later French psychologist, Pierre Janet, writing in 1928 rejected the idea that we have any specific sense of time. More recently another distinguished psychologist in Paris, Paul Fraisse, has criticized Janet for going too far. In his view, although Janet was correct in his claim that our feelings of duration are reactions to the nature of our actions, he overlooked the fact that some of our actions – such as the synchronization of our movements to periodic stimuli when we dance – are direct adaptations to time.

The problem puzzled Robert Hooke in the seventeenth century. He wrote:

> I would query by what sense it is we come to be informed of Time; for all the information we have from the senses are momentary, and only last during the impressions made by the object. There is therefore yet wanting a sense to apprehend Time; for such a Notion we have; and yet no one of our Senses, nor all together can furnish us with it and yet we conceive of it as a Quantity. . . . Considering this, I say, we shall find a Necessity of supposing some other Organ to apprehend the Impression that is made of Time. And this I conceive to be no other than that which we generally call Memory, which Memory I suppose to be as much an Organ as the Ear, Eye or Nose, and to have its Situation somewhere near the Place where the Nerves from the other Senses concur and meet.

Hooke argued that what he called 'the Soul', which nowadays we would call the mind, can no more remember without the organ of memory than it can see without the organ of sight. For the soul can only effect its will by means of corporeal organs. He regarded memory as a repository of ideas formed partly by the senses but subject to what he called 'the Directive Power of the Soul'. This action of the soul, he said, is that 'which is commonly called Attention'. He suggested that in the action of attention the soul directly manipulates some material parts of the organ of memory. 'The Soul', he wrote, 'therefore understands Time or becomes sensible of Time only by the Organ of Memory.'

The mechanism of memory is still one of the greatest unsolved problems of the science of time. Two celebrated hypotheses – the prototypes of two different kinds of explanation of memory that have persisted to the present day – are to be found in Plato's dialogue the *Theaetetus*, in which he examined the problem of how we are able to extract knowledge from our sense-experiences. His solution was the theory of *Anamnesis*: that knowledge is acquired by the recollection in this life of eternal realities and truths by the soul prior to its experience of the external material world. In the course of his discussion, Plato was naturally led to consider the nature of memory.

First, he compares the mind to a block of wax and suggests that memories are impressions imprinted upon it. He soon points out, however, the inadequacy of this analogy and its failure to explain mistakes of judgement such as when we see a stranger but imagine that he is someone whom we know, that is, someone whom we are capable of remembering. So Plato turns to a more subtle idea and compares memory to an aviary. This beautiful simile enables him to explain how the image of something may be registered in the memory without our being consciously aware of it. Speaking through the dialogue-character Socrates, he says:

> Once more then, just as a while ago we imagined a sort of waxen block in our minds, so now let us suppose that every mind contains a kind of aviary stocked with birds of every sort, some in flocks apart from the rest, some in small groups, and some solitary, flying in any direction among them all. When we are babies we must suppose this receptacle empty, and take the birds to stand for pieces of knowledge. Whenever a person acquires any piece of knowledge and shuts it up in his enclosure, we must say he has learnt or discovered the thing of which this is the knowledge, and that is what 'knowing' means.

Socrates continues:

> Now think of him hunting once more for any piece of know-
> ledge that he wants, catching and holding it, and letting it go
> again. In the same way, if you have long possessed pieces of
> knowledge about things you have learnt and know, it is still
> possible to get to know the same things again, by the process
> of recovering the knowledge of some particular thing and
> getting hold of it. It is knowledge you have possessed for some
> time, but you had not got it handy in your mind.

The two analogies suggested to Theaetetus by Socrates in this
dialogue are themselves analogous to the modern distinction
between the idea of a passive memory mechanism and more
dynamic explanations of how memories are stored and recalled. It
was Sir Frederic Bartlett, in his book *Remembering* published about
forty years ago, who first showed conclusively by concrete examples
that both the subconscious retention of memories and their conscious recall depend on dynamic factors – that is, to revert to the
Socratic analogies, our memory-mechanism is more like an aviary
than a wax tablet. He showed that 'long-distance' remembering is
not a simple re-excitation of innumerable fixed traces but is essentially an imaginative reconstruction depending on our frame of mind
at the time of recall and using only a few striking details which are
actually remembered, as often happens when we repeat a story that
we have heard or read. Now, it is a common observation that, in
general, we recall best those thoughts that are connected with our
special interests. Thus, a man who is innocent of other intellectual
accomplishments may have a phenomenal knowledge of statistics
relating to certain games and sports. This is because he is constantly
going over these things in his mind, so that they are for him not just
so many odd facts but a concept system full of inter-relations, every
fact being retained by the combined suggestive power of the whole
mass.

The importance of the associations and setting of our individual
memory-elements can hardly be exaggerated. If we recall a past

event without any associations or setting, it is usually very hard for us to decide whether it is an act of our memory or of our imagination. On the other hand, if we entertain an imaginary proposition for a long time and constantly refer to it we may eventually convince ourselves that it is a true memory, just as George IV did when in later life he actually believed that he had been present at the Battle of Waterloo and had led a cavalry charge.

In old age one's clear memories of childhood often contrast painfully with one's inability to remember what happened five minutes ago. But even something that we would not expect to be able to forget sometimes eludes us in later years. Towards the close of his life the famous Swedish botanist Linnaeus took great pleasure in reading his own books and would cry out: 'How beautiful! What I would not give to have written that!'

The usefulness of memory is so obvious that we tend to regard forgetting as a defect. Nevertheless, annoying though it often is to us, our ability to forget is no less valuable than our ability to remember. For our minds certainly retain below the level of consciousness a vast mass of memories that we normally never recall and never need to recall, and it is even possible – although the evidence is not conclusive – that we actually retain a record of everything to which we have at some time paid attention. This was suspected long ago before modern scientific evidence had been obtained – for example, by Diderot. 'I am led to believe', he wrote,

> that everything we have seen, known, perceived, heard – even the trees of a deep forest – nay, even the disposition of the branches, the form of the leaves and the variety of the colours, the green tints and the light; the look of grains of sand at the edge of the sea, the unevenness of the crests of waves, whether agitated by a light breeze or churned to foam by a storm; the multitude of human voices, of animal cries, and physical sounds, the melody and harmony of all songs, of all pieces of music, of all the concerts we have listened to, *all of it, unknown to us, exists within us*. I actually see once more,

wide awake, all the forests of Westphalia, Prussia, Saxony and Poland. I see them again in a dream, as brightly coloured as they would be in a painting by Vernet. Sleep has taken me back to concerts as freshly performed as when I attended them. Dramatic productions, comic and tragic, come back to me after thirty years; the same actors, the same pit . . .

More than a century after Diderot, Bergson and Freud developed theories of memory. Despite their differences, both agreed that *all* forgetting is due to failure of our powers of recall rather than of retention. In other words, forgetting applies only to the conscious mind and below the level of consciousness all memories persist.

This conclusion is confirmed by hypnosis and experiments in brain surgery. The most striking evidence has been obtained by the Canadian brain surgeon Wilder Penfield from electrical stimulation of the cortex. In operations on patients suffering from focal epilepsy he found that the application of a stimulating electrode to the cortex of the dominant lobe induced them to recall specific memories of earlier experiences. These 'flashbacks', as they are called, are usually of utterly unimportant incidents that the patient would never recall voluntarily. The particular experiences elicited presumably depend on chance, but the same 'strip of time' tends to be reactivated on subsequent stimulation. Penfield has argued that these involuntary flashbacks are not examples of memory as we usually understand the term, although closely related to it. No man can recall by voluntary effort the wealth of detail revealed by flashback. A man can learn a song perfectly but he can seldom recall in detail the many times he heard it. Most things that we recall are generalizations and summaries. Patients say that the experience brought back by the electrode is much more real than remembering and is more like actually living through the past once again. The electrode activates all those things to which he happened to pay attention in the interval of time concerned. But, despite this doubling of consciousness, the patient remains fully aware of the present situation. Indeed, he often cries out in astonishment that he is hearing and seeing friends who he knows are, in fact, far away, or even no longer alive.

Penfield's discovery of flashback at first led him to claim that it implied a more or less precise location of the memory trace, but the criticism was soon made that memories are not necessarily stored in those parts of the brain from which they can be elicited. On the contrary, K. S. Lashley found that major cortical excisions in animals such as apes and rats often made very little difference to memories acquired by learning. Lashley concluded that memories do not depend on *localized* engrams, or memory-traces, but on factors affecting the cortex, or a particular region, as a whole.

Although we may accept Lashley's general conclusion that long-term memory-traces are not precisely localized in the cortex, it is difficult to accept his specific hypothesis that these engrams are more or less stable resonance patterns of electro-chemical vibration of nerve cells over comparatively large areas. For it is difficult to reconcile this idea with the well-known fact that, although after severe shock there is usually no recollection of immediately preceding events, long-term memories are remarkably durable and can even be retained after epileptic fits which convulse the brain with electrical activity. Moreover, in animals learned patterns of behaviour survive hibernation when there is practically no electrical activity of the brain at all.

There is some physiological evidence to support the view that long-term and short-term memories are stored differently; the latter probably being located in the hippocampus, two elongated prominences just below the cerebral cortex, whereas it is thought by many neurophysiologists that long-term memories are stored in the cortex itself. Moreover, short-term or immediate memories of a few seconds' duration may be associated with electrical pulses circulating in closed groups of nerve-cells. The failure of this type of explanation in terms of electrical connections *between* cells to account for long-term memories has led to the search being concentrated on chemical processes *within* cells or in their connections with other cells.

The most exciting experiments that have been devised to test the chemical theory of memory have been 'transfer' experiments with planaria, or flatworms. These primitive creatures, not more than half an inch long, are capable of learning a conditioned response of the type studied in dogs by Pavlov. But, unlike dogs, if cut in half they will regenerate, the head growing a new tail and the tail a new head. Normally a flatworm responds to a light source by reaching towards it, but when it receives an electric shock it curls up. Given both light and shock simultaneously it responds to the latter and can eventually be trained to curl up even when the light is turned on. When a flatworm conditioned in this way is cut in half, not only does the new worm generated from the old head remember that it has learnt to contract in the presence of light, but it has been claimed that the worm with a new head grown from the tail of the original worm remembers this too. It was therefore suggested that some chemical memory trace migrates through the primitive nervous system of these creatures.

An even more astonishing claim, first made a few years ago in the United States, was that if the original trained worms are chopped up and fed to untrained ones the latter absorb the learned behaviour along with their food. Perhaps it is just as well that it is impossible for human beings to learn in this cannibalistic manner! For, unlike flat-worms, we humans cannot absorb the type of giant molecules believed to act as chemical messengers without breaking them down into their components.

In none of these experiments is there yet any firm evidence that *memory* has really been transferred from one animal to another and not just some substance that speeds up the learning process. It is, in any case, a far cry from the simple type of conditioned response learning in flatworms to the long-term retention of specific events that characterizes human memory. The human mind not only stores a formidable mass of detail but can reproduce a sequence of past events in chronological order, as has been found in flashback. Although automatic and conditioned reflexes may be maintained chemically or mechanically and immediate memories may be maintained by processes similar to those in large electronic computers, the problem of the mechanism of human long-term memory is still a complete mystery.

As we saw in Chapter 1, we have good reason to believe that all animals except man live in a continual present. Memory has long been regarded as the concomitant of our sense of personal identity. It is the means by which the record of our vanished past survives within us and is the basis of our consciousness of self. If, however, our entire past awareness or a large part of it survives within us unconsciously, why is it that we can remember nothing whatsoever of the events of early childhood? The most convincing answer to this question is that childhood amnesia is due to a time-lag in the development of the conceptual and conventional mental apparatus that we learn to construct as a means of recalling the past.

Unlike animals, man has a sense not only of the past but also of the future, and it was argued by Guyau in his essay on the origin of our idea of time that the original source of this idea was an accumulation of sensations that produced a mental perspective directed towards the future. This hypothesis is supported by the current opinion of anthropologists that the original development of the prefrontal lobes of our brains may have been intimately associated with the growth of our powers of adjustment to future events. For although Neanderthal Man may have shown some rudimentary concern for the future, since he appears to have buried his dead, the emergence of *Homo sapiens* has been correlated with a strongly increased tendency to look forward, the principal evidence being the sudden development of tools which, unlike the primitive Neanderthal hand-axes, were used to make a variety of other tools, such as barbed harpoons, fish hooks and eyed needles, for future use. Incidentally, patients whose prefrontal lobes have been removed surgically tend afterwards to be restricted to the present.

There is no evidence that animals have any more sense of the future than they have of the past. The strongest candidate for being an exception to this rule is the chimpanzee. The problem of whether chimpanzees have any awareness of the future has been studied very carefully by Wolfgang Köhler. In his book *The Mentality of Apes* he explains how he came to the conclusion that their behaviour, even if it seemed to show some signs of thought for the future, could best be explained in other ways. For example, experiments in which chimpanzees readily responded to the opportunity given them to postpone eating until they had accumulated a goodly supply of food to eat later in some quiet corner free from disturbance by their greedy companions were thought by Köhler to indicate fear of competition from others rather than any consideration of the future.

It appears then that man is distinguished from all other animals by his sense of past and future – that is to say, by his consciousness of time. Like our other mental abilities, we must regard our awareness of time not as simply a question of sensory response, like taste or smell, but as a potential capacity which we can only realize in practice by learning how to develop it on the basis of our own experience. Thus our perception of change and hence of temporal succession involves acts of mental organization. We may perceive change with all our senses, but they are not homogeneous in this respect, our most sensitive organ for temporal discrimination being the ear. The shortest perceptible duration for vision is about two-hundredths of a second – the principle of the cinematograph depending on a succession of pictures merging if shown more rapidly than that. But the lower limit for *conscious* auditory experience is about two-thousandths of a second. As for *subconscious* auditory discrimination, our powers are amazing. It has been found experimentally that our cues for sound localization are provided by the difference in time for sounds to arrive at our two ears. When sound originates within a few degrees of the plane dividing the head into left and right halves, this time difference may be only about a fortieth of a millisecond (thousandth of a second). Nevertheless, it still provides an effective cue.

Our awareness of the sequence of events depends on the sense with which we perceive them. We find it difficult to place a sensation of one kind (say, visual) between two of another kind (say, auditory) if the latter are in too close succession. But if we had some direct perception of time itself, then the nature of the particular sensations determining the intervals concerned would be of no particular significance. Thus it is not time itself but what goes on in time that produces effects. Time is not a simple sensation but depends on processes of mental organization uniting thought and action.

Light has been thrown on the nature of the mental activity involved in our construction of a 'sense' of time by the study of a bizarre form of amnesia known as the Korsakoff syndrome after the first psychiatrist to give a detailed clinical account of it in 1890. In this illness the patient appears otherwise perfectly normal, with his customary powers of perception, but he seldom if ever speaks about the present and the immediate past. He often retains a good memory for remote events, but when questioned remembers hardly anything about recent events. Dr George Talland of Massachusetts General Hospital, in his book *Deranged Memory*, has described a particular Korsakoff patient whom he calls Helen. After many years in hospital Helen consistently maintained that she had been admitted only the day before. She presented herself as a lady of leisure living in one of the better residential hotels of the town, to which she always expected to return the same day. Although this was probably how she had lived at one time, she seemed quite unaware that the conditions of that highly agreeable life had long since ceased. 'One would think', writes Dr Talland, 'that an elaborate façade such as hers could not be maintained without immense effort and something very much like rigorous mental drill, for she did not once betray her disbelief in anything she pretended. Helen was our one Korsakoff patient who responded to hypnosis but repeated attempts with that technique produced absolutely no new material of recall, nor an admission of an awareness of her condition.'

How can one explain patients like Dr Talland's Helen? Various interpretations of the syndrome have been advanced, such as a disturbance of temporal recording of memories, and an inability to comprehend the flow of time. Korsakoff patients register events but cannot correlate them with subsequent relevant information. They tend to spend long intervals of time doing absolutely nothing, and it is well known that the temporal distance of a past event is easily underestimated if the intervening range is largely empty. The temporal frame of reference becomes distorted and with it the individual's idea of himself. The concept of self is based on our recollection of experiences in objective time that have been organized by the mind into some kind of conceptual structure. Korsakoff patients cannot relate their current experiences to this structure.

A similar failure occurs in the phenomenon of *déjà vu*. This is the sense of false familiarity that occasionally characterizes our awareness of the present so that we have the uncanny feeling that we have experienced long ago what is actually happening now. A good description of *déjà vu* is given by Dickens in *David Copperfield*. In Chapter 39 he writes:

> We all have some experience of a feeling, that comes over us occasionally, of what we are saying and doing having been said and done before, in a remote time – of our having been surrounded dim ages ago, by the same faces, objects and circumstances – of our knowing perfectly what will be said next, as if we suddenly remembered it!

This type of experience may have been one of the psychological sources that gave rise in antiquity to the doctrine of metempsychosis, or the migration of souls, held by the Pythagoreans and others. According to some neurologists, the part of the brain that is involved in *déjà vu* is probably the hippocampus. Be that as it may, it has been found that patients who have had this region removed surgically tend to forget the incidents of daily life as quickly as they occur, although they clearly remember incidents of their childhood. It may well be a lesion in this part of the brain that gives rise to the Korsakoff syndrome.

We have already seen that nowadays our sense of time is believed to have originated as a product of human evolution, and that our perception of temporal phenomena is not regarded as a purely automatic process, as Kant thought, but as a complex activity which we develop by learning. Our conscious awareness of time depends on the fact that our minds operate by *successive* acts of attention. For it appears that we cannot attend to two simultaneous events and perceive both of them clearly unless they are combined in some way. The problem of how time-order is constructed by us on the basis of our movements of attention is, however, far from simple. Our memories can be notoriously unreliable guides to the order of events as they actually happened. The fact that patients in hypnotic trance have been found to possess a far more accurate sense of time than in their normal state not only confirms the existence in ourselves of permanent organic rhythms but also indicates that in the normal functioning of consciousness all such rhythms are overshadowed by the transience of external events.

Despite permanent organic rhythms in our brains and bodies, our sense of temporal duration is often wildly inaccurate. It depends on how preoccupied we are and whether we are actually experiencing the time in question or looking back on it. As Thomas Mann stressed in his novel *The Magic Mountain*, the greater the interest we take in what we are doing the faster the time seems to pass; but when we look back upon this time, the fuller its content the longer it appears to have been. A tedious hour is felt to be endless while it lasts, but long intervals of boredom tend to shrink together in one's memory. Complete uniformity would make the longest life seem short whereas the eventful years 'flow more slowly than the poor, bare, empty ones over which the wind passes and they are gone'.

Our sense of temporal duration also depends on our age, for our organic processes tend to slow down as we grow older, so that, compared with them, physical time appears to go faster. This apparent speeding-up of physical time with age is the theme of the verse by Guy Pentreath:

> For when I was a babe and wept and slept,
> 　Time crept;
> When I was a boy and laughed and talked,
> 　Time walked;
> Then when the years saw me a man,
> 　Time ran,
> But as I older grew, Time flew.

Our emotional life also influences our experience of time. Temperamental differences between those who cannot delay communicating or acting upon their thoughts and those who only do so, if at all, after much delay and prompting, can be regarded as variations of time-experiencing. This is borne out by comparing the time concept of normal adults, reared in the traditions of Western civilization, with the very different ideas of time entertained by people of other cultures. Today we tend to regard our recognition of the linear progressive nature of time as instinctive and inevitable. No doubt this idea is influenced by the fact that the process of thinking has the form of a linear sequence. Nevertheless, our ability to synthesize into a single unidimensional time-order the experiences associated with our different senses is a sophisticated late product not only of our biological but also of our social evolution.

3

Biological clocks

ESPITE THE MANY UNSOLVED PROBLEMS concerning our conscious awareness of time, there is increasing evidence that our bodies contain various biological clocks. For example, some illnesses reveal diurnal, or twenty-four-hour, rhythms which normally cannot be detected. Dr C. P. Richter of Baltimore has cited the case of a bedridden patient, unable to speak clearly, who every day for nine years underwent a sudden change of personality between 9 p.m. and midnight, during which time she could walk about, look after herself and talk clearly. He has suggested that there are a number of clocks in the body which normally are not correlated but which under certain abnormal conditions are brought into phase so as to produce well-defined physical or mental cycles.

Although it has long been known that animals and plants show daily and seasonal rhythms and that certain patterns of behaviour occur periodically, it has become evident only in recent years that these phenomena depend on an inner capacity for the measurement of time by living organisms.

In the case of man, the faculty of physiological time-measurement often enables one to estimate with a high degree of accuracy periods of several hours in the absence of external indicators. This is noticeably the case with those of us who are able to wake up at a given time without the aid of an alarm clock. This 'head clock', as it is sometimes called, works with most precision under hypnosis: an order to perform a particular action after a stated time interval will usually be obeyed with remarkable accuracy. Even without hypnosis, the 'head clock' can often function with considerable precision over fairly long periods. In a famous experiment in 1936 two subjects shut up in a sound-proof room for forty-eight hours and eighty-six hours, respectively, estimated the time with such accuracy that their respective errors amounted to less than 1 per cent. On the other hand, two men who spent nearly five months underground in separate caves in 1968–9 with no means of telling the time were found after two weeks to have greatly underestimated its passage, and their reckoning of a day changed from twenty-four hours to forty-eight. The need for some cues in order to estimate the length of the day, as revealed by this experiment, merely shows that under such abnormal conditions psychological influences eventually predominate over physiological factors. This does not invalidate the hypothesis of the biological clock.

Animal navigation

Evidence for the existence of internal timing mechanisms in animals and plants has come mainly from three different fields of research: the study of animal navigation, the study of photoperiodism (the general name given to the responses of living organisms to seasonal changes in the lengths of day and night), and the study of daily and other periodic rhythms in the behaviour and activity of living organisms.

It has long been known that migrant birds can fly great distances to specific destinations, and that even quite young birds can make their way independently of adults of the same species. But it was not until 1949 that Gustav Kramer discovered the explanation. He observed that starlings confined in a cage out of doors at migration time indicated by their behaviour the direction in which they wished to travel. They tended to head in a particular direction whether they were hopping about or sitting on a perch fluttering their wings. Kramer noticed, however, that they did not do this when the skies were completely overcast. On shielding the birds from the direct rays of the Sun and using appropriate mirrors, he found that the birds' sense of direction depended on the apparent position of the Sun. He also discovered that, if the birds were kept in an enclosure illuminated by an artificial sun that was kept in a fixed position, they systematically changed their orientation during the day at a rate corresponding to the Earth's rotation. He trained starlings to feed in a given compass direction at a particular time of day and then tested the birds at another time. He found that they still took up the training position. The birds obtained their compass direction by sighting on the Sun and making due allowance for its regular daily motion. Kramer also discovered that, if a starling trained in this way were confined in a cage with an artificial day and night six hours out of phase with the natural day and night and were later placed in natural sunlight, it would seek food in a direction ninety degrees from the true direction. This suggested that the internal clock used by these birds to obtain direction from the Sun was maintained by the daily light–dark cycle of the locality and hence in this case was six hours out of phase with natural time.

Similar experiments have since been made by E. F. G. Sauer on birds that migrate mainly by night. He confined some migrating warblers in a sound-proof cage within a planetarium. Although they were given no external indication of the time of year, when autumn came they began to flit about restlessly, night after night, as if informed by an internal clock that it was time to take wing. When the appropriate star pattern was cast on the ceiling they indicated their direction of migration. No particular stars or constellations seemed to be involved in their direction-finding ability but only the general pattern of the night sky. Since they made due allowance for the apparent rotation of the sky during the course of the night, Sauer concluded that these birds navigate with the aid of an internal clock that allows them to relate the appearance of the heavens at each season to terrestrial geography. It is now thought that, although the stars are not used for navigation, direction is obtained at sunset and maintained through the night with their aid.

A biological clock seems also to be involved in the homing instinct of pigeons. Although these birds were used by the ancient Egyptians, Greeks and Romans to convey messages, no other use was made of their homing abilities until pigeon racing was introduced by the Belgians in 1825. This sport depends on the fact that not all pigeons are equally proficient in homing, only a few per cent being able to make long-distance returns at high speed. Experiments have shown that if pigeons are exposed to an artificial day and night out of phase with local time and are then released some way from home they will usually head in the wrong direction; nevertheless they often reach home eventually. Although it therefore seems probable that they have some awareness of topographical features, they must also have an accurate biological clock.

Migrating birds and pigeons are not the only animals that use an internal clock for direction-finding. The sand hopper (*Talitrus saltator*) does too. This animal inhabits the wet sand of beaches and if removed to dry sand tries to escape back to the sea in a direction at right angles to the shore line. It determines the required direction by means of the Sun's position, and for this it relies on an internal clock. The existence of such a clock was established by experiments, similar to those on starlings, which showed that the clock can be shifted by shifting the light–dark cycle.

Perhaps the most sophisticated application of an innate time sense to direction-finding is that made by bees. Their agility, when laden with food, to fly directly back to the hive has long been known. It has given rise to the term 'bee-line' for a straight line joining two places. Their ability to tell the time, however, has only been realized in this century. It was first studied by a Swiss doctor, Auguste Forel, who was in the habit of breakfasting at the same time most mornings on the porch of his house. Bees used to come and take small portions of the jam left over at the end of the meal. Forel observed that they arrived at the same time each morning, even on those days when he did not breakfast out of doors. Although it had previously been known that bees come to certain flowers only at those times of day when they secrete nectar, no one had studied this. It is only since the last war that the time-sense has been systematically investigated, notably by Karl von Frisch and his colleagues in the University of Munich. They found that, although bees could be trained to come daily to a particular feeding station at the same time, they could not be trained to come to the same place at different times. On the other hand, they could be trained to feed at two different places at two different times of day or even at several different places at several different times. Frisch concluded that the time-sense of bees is not based on the learning of intervals but depends on an internal clock with a period of twenty-four hours. This was confirmed by training bees to feed at a certain time of the day in a European time zone and then transporting them by plane to an American time zone where it was found that their behaviour continued in phase with European time.

The remarkable ability of bees to communicate with one another also depends on the use of a biological clock. When a scout bee finds a bed of nectar-laden flowers he hastens to inform the other bees in his colony. On returning to the hive he performs a dance indicating the distance and direction of the flowers. If the flowers are near by – not more than fifty to a hundred yards off – the scout bee performs what Frisch has called a round dance, turning round once to the right and once to the left and repeating these circles with great vigour for half a minute or more. But if the flowers are some way off, the dance is quite different. The bee runs a short distance wagging its abdomen rapidly from side to side. It then makes a complete turn leftwards, runs forward again in the same direction as before still wagging its abdomen and then turns to the right and repeats the pattern over and over again. This wagging dance not only announces that the source is some way off but also provides information about its distance and direction.

Distance is indicated by the number of turns made in a given time: the smaller the number the greater the distance. Although the way in which distance and rate of turn are related shows some variation from one colony to another, there is little variation inside a given colony from one bee to another. Strictly speaking, however, the rate of turn is not always a true indication of distance, for it also depends

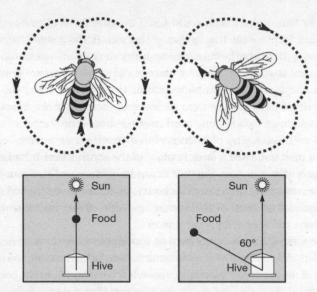

Figure 2: The dance of the bees. In the example on the left the bee is indicating flowers some way off, in the same direction as the sun; vertically up the face of the comb means 'towards the sun'; changing the angle of the run to left or right of the vertical means a similar change in the directional information. The bee gives an idea of the distance by the speed of the dance: the faster this is performed, the nearer are the flowers.

on the direction of the wind. A head wind on the way to the feeding place has the same effect as increased distance: it slows the dances. On the other hand, a tail wind has the opposite effect. The basis of the bee's estimate of distance thus seems to be the time or effort needed to reach the flowers.

The direction of the flowers is indicated by the straight part of the dance, the Sun being used by the bee as a kind of navigating compass. For it has been found that the direction of the dances relating to the same flowers changes during the course of the day by approximately the same angle as the Sun. Frisch and his colleagues were surprised to discover that even when part of the sky is cloudy, and the Sun is obscured, bees can still indicate the correct direction of a feeding place relative to that of the Sun. It was found that they do this by utilizing the relationship between the Sun's position and the polarization of the blue sky. More remarkably, if bees are induced to dance during the night, they indicate the direction of the place at which their daily feeding time is closest to the time of the dance. In one experiment a foraging bee announced at 9.31 p.m. a feeding place in the east where it had been fed every day at 6 p.m., but at 3.54 a.m. it indicated a feeding place in the south where it had been fed every day at 8 a.m. Further experiments revealed that not only can bees store in their memories both feeding times and feeding sites but also the azimuth of the Sun at any time of the day even when they have not seen it for several weeks.

There are strong indications that knowledge of the Sun's orientation and of the fact that it moves is not known to the bees innately but has to be learned. A similar conclusion has been drawn in the case of birds. K. Hoffmann took six young starlings from their nest box and reared them without any direct view of the Sun. They were then direction-trained at a particular time of day with the aid of an artificial sun, but when tested at other times only two managed to make some allowance for the direction of the Sun's motion and then not accurately for its speed. The others maintained the original training angle.

To sum up: the various phenomena revealed by the navigational achievements of animals cannot be understood unless we assume that they possess some form of internal time-keeping mechanism which they can learn to use. There must be some rhythmic processes going on in them which can serve as reliable clocks. Although these rhythms can continue in the absence of external changes, their pace is maintained by external rhythmic events of which the light–dark succession of day and night is one of the most important.

The stimulus of daylight

In the case of plants, the influence of the relative length of day and night on flowering response was first studied systematically by W. W. Garner and H. A. Allard who, in 1920, introduced the term *photoperiodism*. The concept was soon extended to explain other phenomena of plant and animal behaviour. It had, of course, long been realized that the budding of plants, arousal of animals from hibernation, the seasons for animal breeding and for the formation of winter pelts, as well as the time of migration of birds, were all dependent on environmental changes, but it had not been generally appreciated that the length of daylight was the crucial environmental factor involved. The use of this factor by living organisms is, however, natural, at least in middle and high latitudes, since this is the most regular seasonal change that occurs in these regions of the Earth's surface. Its use no doubt became important during the evolutionary conquest of dry land, where temperature is a less reliable indicator of seasonal change than in the sea.

Although its scientific investigation is modern, photoperiodism in birds has long been exploited by man for purely practical purposes. The Japanese caused male songbirds to sing out of season by keeping them in artificial conditions of lighting in winter. In the Middle Ages, Dutch bird-catchers provided artificial light at night to induce breeding characteristics of song and display in certain birds in autumn so as to use them as decoys for passing migrants. By the opposite process of artificially decreasing the period of daylight, birds were induced to speed up their deposit of fat reserves in readiness for winter and in this way were made ready for the table.

The first conclusive proof that the length of daylight could influence the reproductive behaviour of animals was given by the Canadian zoologist William Rowan, in the late 1920s. He studied a bird known as the greater yellowleg, which migrates to Patagonia in the autumn and returns to its breeding grounds in Canada every spring. Although the round trip is about 16,000 miles, the precision of the bird's timing is such that its eggs are always hatched between 26

May and 29 May. Rowan studied this bird for fourteen years and considered all possible factors that might be involved. He concluded that the only one that was sufficiently regular and precise to act as the required synchronizer was the variation in the length of daylight. To test this conclusion he took birds of another species which normally winter in Canada and subjected them to artificially increased day lengths. He found that when, after a few weeks, they were experiencing daylight conditions that normally were not present until late spring, they were already in the condition to breed, whereas control specimens kept under natural winter-light conditions were not. He studied the effect of releasing birds at various stages of their reproductive development and found that they migrate when passing from the inactive to the full breeding state. From these researches, and others made since, it is clear that photoperiodism is a crucial factor in the reproductive cycles of many species of animal.

In plants the organs which receive the photoperiodic stimulus are the leaves. On being exposed to some critical length of daylight, they transmit a message to the buds causing them to form flowers instead of leafy shoots. Not all flowering, however, depends on day length, for the photoperiodic control of developmental processes tends to occur in those parts of the world where it is likely to be most effective.

As already mentioned, photoperiodic control is found most frequently in organisms living in those temperate middle latitudes where temperature is not a reliable criterion of seasonal change. It is therefore significant that the timings involved in photoperiodic reactions tend, within reasonably wide limits, to be independent of temperature. A good example is the rhythm of expansion and contraction of the pigment cells of the common fiddler crab. This animal shows a twenty-four-hour cycle of variation in colour. During daylight the black pigment in its skin cells disperses through them and makes the crab dark, so as to protect it from the sun and from predators. At nightfall it becomes paler and at dawn the cycle begins all over again. When some of these crabs were placed in a dark room that was maintained at a uniform temperature it was found that different temperatures between 26°C and 6°C had no appreciable effect on this cycle. Although at the lower temperature the actual expansion of the pigment cells was reduced, the timing of the cycle was maintained to within a few minutes over an interval of two months. When the temperature was lowered to 0°C the rhythm disappeared, but was restored when, later, the temperature was raised, although it was out of phase by the appropriate amount. For example, if the lower temperature was maintained for six hours the restored rhythm was out of phase by a quarter of a cycle, whereas if the low temperature continued for twenty-four hours the restored rhythm was in phase. Similar independence of temperature is found for photoperiodism in plants.

So far we have spoken of the influence of length of daylight, but strictly speaking it is not the length of day as such that is decisive but rather the length of night. This was discovered by observing that the effects of a long period of daylight could be obtained after a short day if the dark period were interrupted by a comparatively short spell of light. The time of highest sensitivity to this interruption is not necessarily in the middle of the dark period but often occurs a certain number of hours after its onset. For example, although breeding can be initiated in ferrets within two months when they are subjected to eighteen hours of continuous light daily, a six-hour daily total is just as effective if two hours of light are given as an interruption of the dark period twelve hours after it begins. In some cases, however, the beginning of the light period has more influence on the timing of the critical point of highest sensitivity during darkness than does the beginning of the dark period itself. The increasing and decreasing sensitivity to interruptions during the dark period suggests that the physiological process concerned is controlled by some internal clock.

Clearly there is a great deal of evidence for believing that living organisms possess internal biological clocks which enable them to make surprisingly accurate time measurements. Further confirmation of this conclusion has been provided by the discovery that many organisms show cyclic behaviour even when no change occurs in their physical environment. This appears to have been first observed by a French astronomer, Jean Baptiste de Mairan, in 1729. He was interested in the leaf movements of plants. Many plants extend their leaves in the hours of daylight and fold them at night. Mairan found that these movements will continue even if the plants are kept in constant darkness. Later, other scientists studied similar phenomena and Darwin discussed them in his book *The Power of Movement in Plants*. Such movements were regarded as being due either to an after-effect of exposure to the cycle of day and night or else to some inherited tendency for movement. It was not until the 1930s, mainly through the researches of Erwin Bünning of Tübingen, that it came to be realized that these daily rhythms are evidence for the existence of intrinsic biological clocks and that plants can accurately measure the passage of time, even when placed in complete darkness.

Bünning also found that the cycle of leaf movement had a period that was only *approximately* twenty-four hours. This was a crucial discovery, for it clinched the argument that the rhythmic movements were not controlled by an external factor but were intrinsic to the plant itself and must therefore depend on an internal clock. In recent years the term 'circadian', from the Latin *circa diem* meaning 'about a day', has come into general use to denote all biological rhythms that deviate somewhat from precise twenty-four-hour periodicity. A rhythm that was not circadian but actually equal to the length of the day would have to be attributed to some geophysical agent. A circadian rhythm, on the other hand, although it presumably had its evolutionary origin in conditions dependent on the length of the day, maintains a different periodicity. This is a strong indication that the rhythm is an intrinsic characteristic of the organism itself, particularly since it does not synchronize with any known daily change in the environment.

We now know that circadian rhythms are exhibited by almost all plants and animals, from unicellular algae to man. To investigate the existence of these rhythms in man, a small group of scientists went to Spitsbergen, where the sun shines continually for several months in the summer and there is little daily variation in light or temperature. They experimented by setting their clocks to complete a twenty-four-hour cycle first in twenty-one hours and later in twenty-seven hours. They carried on all their activities according to these clocks and samples of their urine were collected at regular intervals and analysed. On the whole it was found that their circadian rhythms were maintained.

Not all biological rhythms that have external correlates are circadian. Some marine organisms show rhythms in their behaviour that are clearly associated with high and low tide. For example, green flatworms (*Convoluta*) come to the surface of the sand at high tide and then bury themselves in the sand as it dries. It has been found that this rhythm continues when they are placed in an aquarium where there are no tides. Lunar cycles also occur in some marine organisms. The palolo-worm, found in the Pacific and Atlantic Oceans, reproduces only during the neap tides of the last quarter Moon in October or November. Also the brown alga (*Dictyota dichotoma*) liberates its male and female gametes at certain localities twice in a lunar cycle, thereby increasing the chances of fertilization. Since a number of species continue their periodic behaviour under laboratory conditions without being exposed either to tidal action or to moonlight, it is clear that they must possess intrinsic rhythms with the corresponding periods. Lunar and semilunar cycles may be explicable in terms of the interaction of circadian rhythms with tidal rhythms, certain phases of the two coinciding so as to produce regular beats at intervals of about fifteen or twenty-nine days.

The cycle of the year

The existence of circadian rhythms throughout the animal and plant world has led biologists to inquire whether there are any intrinsic rhythms with a period of about a year. This is more difficult to determine since it clearly requires a lengthy year-by-year investigation. The first positive evidence for an annual biological rhythm was discovered by K. C. Fisher and E. T. Pengelley at the University of Toronto in connection with animal hibernation. A species of ground squirrel that inhabits the Rocky Mountains was kept in a small windowless room at freezing point (0° Centigrade) and supplied with ample food and water. From August until October it ate and drank normally and maintained a body temperature of 37°C despite its cold surroundings. In October, as would happen if it had been in its natural surroundings, it stopped eating and drinking and went into hibernation, its body temperature falling to just above freezing point. In April it became active again and resumed its normal behaviour and body temperature. Similar experiments with other ambient temperatures gave the same result, the period of each complete cycle being a little less than a year. The usual criteria for the existence of an intrinsic biological clock were satisfied: the period of the rhythm was not precisely a year, it did not synchronize with any periodic external signal, and it was independent of the temperature.

The 'circannual' rhythm, as it is now called, was even manifested when these animals were kept at a constant ambient temperature so close to their normal body temperature that it was impossible for them to hibernate. Although food and water were available, they reduced their consumption and lost weight during the winter and then reverted to normal in the spring. As E. T. Pengelley and S. J. Asmundson, who made this experiment, have commented, 'There could hardly be a more convincing demonstration of the existence of an internal clock operating independently of the environmental conditions.'

The discovery of a circannual clock in hibernators has led to the search for similar rhythms in other animals, particularly birds. We have already seen that, by altering the daily rhythm of light and dark, William Rowan influenced the timing of the characteristic restlessness that initiated the birds' migration. Nevertheless, Rowan suspected that the length of day was probably not the only factor that influenced the urge to migrate. His belief has recently been confirmed by the discovery of a circannual clock in some migratory birds. Experiments on warblers, some being kept under natural conditions and others under a constant ambient temperature and a daily cycle of twelve hours' light and twelve hours' darkness, have shown that differences in environmental conditions have little effect on the birds' migratory urge. It did not matter whether they were kept in Europe, where they normally spend the summer, or in Africa, where they normally winter. Eberhard Gwinner of Munich, who conducted these experiments, came to the conclusion that the warblers' rhythm is based on some intrinsic circannual timing mechanism.

Another cycle that has been similarly studied is the annual growth and shedding of antlers by deer. It is well known that tropical deer when confined to zoos in temperate latitudes maintain the same annual cycle despite the difference in the day-length pattern. Impressive evidence for believing that antler growth is controlled by an intrinsic clock is provided by a blind elk that has been under observation for six years at Colorado State University. Despite the absence of any light cues it has shed and regenerated its antlers on time throughout this period.

An intrinsic clock with a period that is not precisely a calendar year has also been found in a species of crayfish that lives in dark caves where there are practically no seasonal changes. Indeed, it now seems probable that circannual clocks are almost as universal as circadian clocks. The advantage of having such a clock is that it gives an animal an advance warning that it may not always get from its environment. Birds wintering in tropical regions near the equator can seldom receive much of a signal from their surroundings to inform them that it is time to migrate to their breeding places in more temperate latitudes.

There is some evidence that circannual clocks exist in man. Long-term studies of a psychotic individual indicated an annual rhythm in his manic-depressive attacks, and in another a similar cycle was revealed in his body weight.

Stress has been laid on the deviations of circadian and circannual rhythms from the exact day and year, respectively. If, however, the external environment played no part in regulating these rhythms, they would become increasingly out of phase with the day–night cycle and with the seasons. Consequently, the animal or plant concerned must depend on some cues or signals from its environment to correct its clock and keep it running more or less in phase, just as we use time signals from our national observatories to regulate our clocks and watches. In the case of circadian rhythms, it seems that the necessary cues are provided by the daily variations in light and temperature. Presumably, these variations are also involved in regulating circannual rhythms, but there may be other influences too.

Master clocks

It is clear that biological clocks of the type discussed cannot depend upon metabolic processes, for their rates would then be temperature-dependent. Of course, metabolic processes can influence the timing of an organism's activities. Bees fed with chemicals which speed up their metabolism tend to arrive too early at the flowers from which they normally obtain nectar. A rise or fall in the inner temperature of an organism can produce an acceleration or a deceleration of its physiological processes, making its metabolic clock beat faster or slower. Thus, it has been found that if flies are kept at an abnormally high temperature they age more rapidly and die sooner.

Although the non-metabolic biological clocks are temperature-independent as far as their rates are concerned, they can nevertheless make use of temperature as a synchronizing clue to adjust their setting and bring it into phase with some significant external condition. Consequently, one of the most important problems now facing biologists is to discover a physiological mechanism that can respond to temperature in the one respect and yet be independent of it in the other. The problem has been studied biochemically, but there is no sign that any enzyme fluctuation is involved in such a clock mechanism. Plants, unlike animals, show no evidence of any central regulator of periodicity. Bünning has therefore argued that they must have a clock in every cell, but although this conclusion is now widely accepted no cellular clock has yet been identified.[1]

The synchronization of a number of cellular clocks could be regulated by some rhythmic control centre. Particular mechanisms that could perform this function have been discovered and are called 'master clocks'. The first such master clock was found by G. P. Wells in the oesophagus of the lugworm (*Arenicola marina*). It controls the extraordinarily regular activity of this animal: three-minute bursts of feeding movement that occur irrespective of the presence of food, followed by one-minute rests and locomotory movements every forty minutes.

Another master clock has since been discovered by Dr Janet Harker of Cambridge in the cockroach (*Periplaneta americana*). If kept in a standard light–darkness cycle, this insect shows a distinct circadian rhythm in its foraging activities, being most active at the onset of darkness. But if it has been kept in continuous light for a long time, it ceases to show any measurable rhythm in its activities. A cockroach with a good rhythm was immobilized by the removal of its legs and was then grafted on to the back of one with no such rhythm but able to move about. The blood systems of the two insects were joined by means of a capillary tube to form a single circulation. Dr Harker found that the lower insect, although still in continuous light, soon developed the same circadian rhythm as had previously been shown by the upper one. Moreover, and this was the crucial finding, the rhythmical cockroach imparted the *phase* of its activity to the other. This was a strong indication that the rhythm is due to the periodic release of some hormone into the blood stream.

By transplantation experiments Dr Harker found that the source of this secretion is the sub-oesophageal ganglion forming the ventral part of the brain. This organ contains specialized nerve cells which secrete hormones under the influence of light entering the eyes of

---

1. C. F. Ehret of the Argonne National Laboratory in the USA has argued that associated with the control system for nucleic acid metabolism there is a cellular clock mechanism.

the insect. When these cells were removed and implanted in the body cavity of another cockroach which had previously had its head removed, they went on functioning as before. The headless insect continued for several days to run round at the time of day to which the phase of secretion of the implanted cells had been set by the donor cockroach's experience of light and darkness.

Further experiments, however, indicate that a second circadian process is involved in the rhythm of the foraging activities of the cockroach. This second clock influences the setting of the master clock (preventing it from being reset by random light flashes such as transitions from shade into sunshine) but is situated outside the sub-oesophageal ganglion. This makes it seem unlikely that there is any simple solution to the problem of the precise nature and location of the complete mechanism controlling circadian and other rhythms in animals.

In the higher animals, including man, the master clock presumably lies in the central nervous system. During the past forty or so years it has been known that the human brain is subject to incessant rhythmical activity due to electric currents. The harmonic analysis of electro-encephalograms is complicated, but four main types of rhythm, each characterized by a particular frequency range, have been recognized. The easiest to detect, particularly when the eyes are shut and the patient is relaxed, is the 'alpha rhythm', of approximately ten cycles a second. It is probable that this rhythm is related to a diminution of information processing, since it is particularly prominent during meditation exercises (Zen and Yoga) and it is well known that the object of these exercises is to isolate the practitioner from all forms of external stimulation.

There is evidence that the predominance of the alpha rhythm when the brain is resting is due to the synchronized fluctuations of large groups of cells, whereas the pattern of electrical activity revealed by the alert brain is of a lower voltage and corresponds to the highly diversified activities of its various parts. Moreover, since we can generate the alpha rhythm artificially by submitting the eye to a visual flicker of about ten a second, it is possible that the natural rhythm is the response of the brain to a flicker due to its own internal oscillations. Mary Brazier and her colleagues in Boston have found that certain local oscillations in the brain have a tendency to pull one another into synchronization. Whatever the brain oscillators may be and however discordant, they tend to produce a comparatively well-timed complex system.

It is only recently that we have begun to understand the true nature and significance of biological rhythms. To function properly an organism must control the timing of its physiological processes and, if need be, prepare itself for certain predictable changes in its environment, such as the different seasons. It therefore requires a sense of time, which we now realize is provided by biological clocks, some of which have external correlates, notably the circadian clocks. A living organism cannot therefore be explained entirely in terms of those physical and chemical concepts suggested by examining it only when dissected, since biological clocks cannot be studied in this way if timekeeping is closely associated with particular organs that can be isolated. In modern biology the *time* aspects of life are becoming more and more important. In particular, their study will help us to understand how we ourselves function. For, although our cognitive time sense appears to be largely controlled by social, psychological and metabolic factors, these are superimposed on the rhythm of the innumerable biological clocks that beat within us far below the level of consciousness.

4

The measurement of time

WE HAVE SEEN THAT the function of biological clocks is to enable living organisms to produce the required responses at the appropriate times. Similarly, the calendars used by previous civilizations and the Gregorian calendar that we still employ were originally devised so that religious ceremonies could be performed on the correct dates. To function effectively, biological clocks (with external correlates) and man-made calendars use cues provided by the motions of the Earth, Sun and Moon. These motions therefore function as time standards. Although, as far as man is concerned, it has long been realized that the correlation of any time-system with a given standard is facilitated by means of measurement, it was only with the rise of modern science that the crucial significance of this became clear. The ideas of time that prevailed in antiquity and the Middle Ages differed from ours not only because of the widespread belief that time is essentially cyclical but also because until about three hundred years ago the lack of reliable mechanical clocks prevented its accurate measurement.

The alternation of day and night, which was responsible for the evolution of circadian clocks in most organisms from unicellular algae to the higher plants and animals, also exerted a widespread influence on the way in which time was measured by man. As a result, the development of our modern concept of the homogeneity of time was impeded, because the scale of hours generally adopted was not uniform. Before the fourteenth century our present system of dividing day and night together into twenty-four hours of equal length was used only by some theoretical astronomers. Instead, the periods of light and darkness were split up into an equal number of 'temporal hours' – *horae temporales*, as they were called by the Romans. The number was usually twelve. Consequently, the length of an hour varied according to the time of year and also, except at the equinoxes, a daylight hour was not equal to a nocturnal hour. Strange as this mode of reckoning time may now seem, it must be remembered that most human activities took place in the hours of daylight and also that early civilizations were in latitudes where the period from sunrise to sunset varies far less than in more northerly parts. On the other hand, astronomers used standard hours, known as 'equinoctial hours' – *horae equinoctales*. These were the same as the temporal hours at the date of the spring equinox.

The only mechanical time-recorders in antiquity were water-clocks, but until the fourteenth century the most reliable way to tell the time was by means of a sundial. Both types of clock were used by the Egyptians. Later they were introduced into Greece and eventually became widespread in the Roman Empire. Vitruvius, writing about 30 BC, described more than a dozen different types of sundial. He also described a number of 'clepsydrae' or water-clocks. To obtain a uniform flow of water they were designed so as to keep the pressure head constant. In order to indicate 'temporal' hours, either the rate of flow or the scale of hours had to be varied according to the time of year. The result was that many of the ancient water-clocks were instruments of considerable complexity.

The earliest known attempt to produce mechanically a periodic standard of time is a device illustrated in a Chinese text written by Su Sung in 1092. It was powered by a water-wheel which advanced in a step-by-step motion, water being poured into a series of cups which emptied (or escaped) every quarter of an hour, when the weight of water in the cup was sufficient to tilt a steelyard. The mechanism was then unlocked until the arrival of the next cup below the water stream, when it was locked again. An astronomical check on time-keeping was made by a sighting tube pointed to a selected star. Since the time-keeping was governed mainly by the flow of water rather

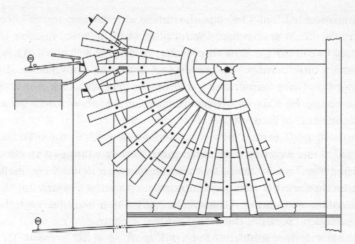

Figure 3: Su Sung's water-clock can be regarded as a very early escapement mechanism. Once every twenty-four seconds the weight of water in one of the cups becomes enough to press it down and trip the counter-weighted horizontal beam at the top. This allows the wheel to revolve by one more spoke, placing an empty cup under the water spout. The loosely hinged beam at top left acts as a ratchet, preventing movement backwards.

than the escapement action itself, this device may be regarded as a link between the time-keeping properties of a steady flow of liquid and those of mechanically produced oscillations.

The fundamental distinction between water-clocks and mechanical clocks, in the strict sense of the term, is that the former involve a continuous process (the flow of water through an orifice) whereas the latter are governed by a mechanical motion which continually repeats itself. The mechanical clock, in this sense, appears to have been a European invention of the later thirteenth century. Already in the late twelfth century the market for water-clocks was such that a guild of clockmakers is known to have existed in Cologne and by 1220 they occupied a special street, the Urlogengasse. Nevertheless, in northern climes water-clocks must have been a nuisance in winter when they froze, and so in the fourteenth century sand-clocks were invented, but they proved suitable only for measuring short periods. They were principally used on board ship to measure its speed by counting the number of knots paid out on a line tied to a log thrown into the water and drifting aft, while the glass measured a period that was usually half a minute. Incidentally, it was not until the end of the fifteenth century that the sand-glass was depicted as the attribute of Father Time.

The English word 'clock' is etymologically related to the French word *cloche*, meaning a bell. Bells played an important part in medieval life and it is probable that mechanisms for ringing them, made of toothed wheels and oscillating levers, prepared the way for the invention of mechanical clocks. The crucial invention that made the mechanical clock possible was the 'verge' escapement (the word was probably derived from the Latin *virga*, meaning a rod or twig). This was an ingenious device in which a heavy bar, or 'foliot', pivoted near its centre, was pushed first one way and then the other by a toothed wheel driven by a weight suspended from a drum. The wheel advanced by the space of one tooth for each to-and-fro oscillation of the bar. In Italy the bar was sometimes replaced by a balance wheel with a similar reciprocating action. No one knows who made this invention, although it must have been towards the end of the thirteenth or early in the fourteenth century. Since the bar had no natural period of its own, the rate of the clock depended on the driving wheel, but it was greatly affected by variations of friction in the driving mechanism. Consequently, the accuracy of these clocks was low and they could not be relied upon to keep more closely than to about a quarter of an hour a day. An error of an hour was not unusual.

Despite their lack of accuracy, many public mechanical clocks that rang the hours were set up in European towns in the fourteenth century. Clocks were made with curious and complicated movements. It was easier to add wheels than to regulate the escapement. Moreover, in view of the general belief that a correct knowledge of the relative positions of the heavenly bodies was necessary for the success of most human activities, many early clocks involved

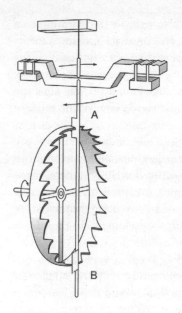

Figure 4:
The ingeniously simple verge escapement. The toothed wheel, driven round by a weight, is prevented from 'running away' by two projections on the vertical rod or verge. In the diagram, the wheel is about to push projection A aside, thus moving the weighted cross-bar through 90° and bringing projection B into play. The verge and cross-bar must now be forced to reverse their swing before the wheel can advance one more notch.

elaborate astronomical representations. The most celebrated was the Strasbourg clock, set up in 1352. From about 1400 there are records of the purchase of domestic clocks by royalty, but until the latter part of the sixteenth century these clocks were very rare.

The invention of the mechanical clock was the decisive step that led to the general use of the system of time-reckoning in which day and night together are divided into twenty-four equal hours. In Italy, where a public clock was set up in Milan in 1335, clocks struck up to twenty-four and this system persisted in that country for several centuries. Most other countries of Western Europe, however, soon adopted the system in which the hours were counted in two sets of twelve from midnight and from noon. In England the first references to hours 'before noon' and 'after noon' occur about 1380.

Until the middle of the seventeenth century mechanical clocks had one hand and the dial was divided only into hours and quarters. Although the division of the hour into sixty minutes and the minute

into sixty seconds was used in 1345, to express the duration of a
lunar eclipse, no actual measurement was involved but only a theoretical computation. (This way of dividing the hour was based on the
sexagesimal system of fractions used by the Hellenistic astronomers
in antiquity and before them by the Babylonians.) Even after the
invention of the mechanical clock, the development of the modern
scientific conception of time was seriously hampered by the lack of
any accurate mechanical means for measuring short intervals.
Thus, in his famous experiments, in the first part of the seventeenth
century, on the rate of fall of balls rolling down an inclined plane
Galileo had to measure time by weighing the quantity of water that
emerged as a thin jet from a vessel with a hole in it. He removed his
thumb from the hole at the start of the experiment and replaced it
when the ball had reached any chosen point.

The origin of modern accurate time-keeping was Galileo's discovery of a natural periodic process that can be repeated indefinitely
and counted – the swinging pendulum. His interest in the pendulum
can be traced back to when, as a medical student at Pisa, he applied
it to the diagnostic description of a patient's pulse. The application
consisted of a board bearing a peg to which was attached a cord with
a bob that could be made to swing. At appropriate places on the
board were written various diagnostic descriptions, such as 'feverish'
and 'sluggish'. The physician had only to control the length of the
swinging cord with his finger to make the swings synchronize with
the patient's pulse and then read off the diagnosis indicated. Later in
life, as a result of much mathematical thinking on experiments with
swinging pendulums, Galileo came to the conclusion that each simple pendulum has its own period of oscillation depending on its
length, and in his old age he contemplated applying the pendulum to
clockwork so as to record mechanically the number of swings. This
step was taken successfully some years afterwards, 1656, by the
Dutch scientist Christiaan Huygens, whose pendulum clock
inaugurated the era of high-precision time-keepers, being the first
reliable machine for measuring physical time with an accuracy of
about ten seconds a day. This measurement could be looked upon

simply as numerical repetition, but it was also a means for the uniform division of a given interval of time – for example, an hour into sixty minutes – and so was analogous to the division of a continuous line of finite length into a number of equal segments. Consequently, the invention of a mechanical clock which could, if properly regulated, tick away continually for years on end greatly influenced belief in the quasi-geometrical homogeneity and continuity of time.

Strictly speaking, Galileo's simple pendulum in which the bob describes circular arcs is not quite isochronous. Huygens discovered that theoretically perfect isochronism (meaning that the period of oscillation is the same for all angles of swing) could be achieved by compelling the bob to describe a special type of geometrical curve known as a cycloidal arc, and this was the essence of his invention. Great as was his achievement, particularly from the point of view of theory, the ultimate practical solution of the problem came only after the invention of a new type of escapement. Huygens's clock incorporated the verge type, but about 1670 a much improved type, the anchor type, was invented that interfered less with the pendulum's free motion.

The accuracy of any mechanical clock not only depends on its construction but has to be checked by constant reference to some natural clock. Throughout history, the ultimate standard of time has been derived from astronomical observations. In due course this led to the hour, minute and second being defined as fractions of the period of one rotation of the Earth on its axis, as determined by careful study of the apparent daily rotation of the celestial sphere. There are, however, different ways of determining this. In ordinary life it is convenient to define time by the position of the Earth relative to the Sun, and the *mean solar day* is the period of one rotation of the Earth relative to the Sun corrected for all known irregularities. Because the Earth's orbit is not quite circular, the relative velocity of the Sun is not strictly uniform. Also, because the Sun's apparent motion in the sky is not along the celestial equator (that is, the projection of the Earth's equator on to the sky), the component of its

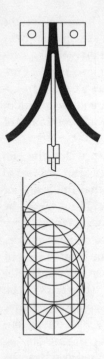

Figure 5:
Confined within metal 'cheeks'
formed in a cycloidal curve, a
pendulum can only swing in such
a way that its bob, too, describes a
cycloidal curve instead of an arc of
a circle. A cycloid is the curve
described by a point on the
circumference of a circle that is
rolled along a straight line (see
diagram). The period of a cycloidal
swing, Huygens discovered, is not
dependent on the angle of swing.

velocity parallel to the equator is variable. Therefore, for the
purposes of time-reckoning a 'mean Sun' is defined which moves at a
rate equal to the average of that of the true Sun. The difference
between mean solar time and apparent solar time, as given by any
form of sundial, is called the 'equation of time'. It may be either
positive or negative, being positive when true noon precedes mean
noon and negative when mean noon precedes true noon. (It vanishes four times in the year: on 15 April, 15 June, 31 August and 24
December.) The mean solar second is defined as 1/86,400 of the
mean solar day.

Despite its convenience in everyday life, solar time is more difficult
to determine accurately than sidereal time, which is based on the
time of transits of stars across the meridian. (The meridian is the
projection on the sky of the circle of longitude through the place on

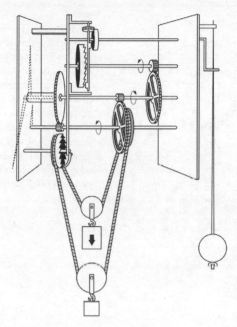

Figure 6: The first clock with a pendulum was built in 1656. This diagram, opened out for clarity, is based on Huygens's own drawing, and shows how the verge escapement (top centre) must have affected the swing of the pendulum to some degree, and thus its accuracy.

the Earth's surface where the measurements are made.) The interval between successive transits of the same star, or group of stars, is the *sidereal day*. Since, owing to the Earth's orbital motion, the Sun appears to us to make one complete revolution eastwards relative to the stars in the course of a year (about 365¼ days), in a day it appears to move eastwards relative to the stars by about one part in 365¼ of a complete circuit (360 degrees), that is by about one degree. It follows that the solar day is about four minutes longer than the sidereal day. The formula for converting from the sidereal to the solar day is 1 mean solar day = 1.0027379093 sidereal days, derived from observations extending over two centuries.

Figure 7: The anchor escapement led to greater accuracy, for the swing of the pendulum was transmitted directly to the anchor. The diagram shows how the pendulum, having swung to the left, forces 'fluke' A between two teeth of the wheel. The swing to the right raises A and the wheel, driven by the weight (not shown), revolves until checked by the descending 'fluke' B.

The precise standardization of time measurement dates from the foundation of the Royal Observatory at Greenwich in 1675. The need for accurate time in those days was primarily felt by navigators. Without an accurate clock that could keep Greenwich time, it was impossible to determine the longitude of a ship at sea, and it often happened that a ship was hundreds of miles off course. The perfection of the marine chronometer by John Harrison, about 1760, was a landmark in the practical standardization of time. Nearly fifty years previously, the British government had passed an Act offering rewards of £10,000, £15,000 and £20,000 to anyone who should

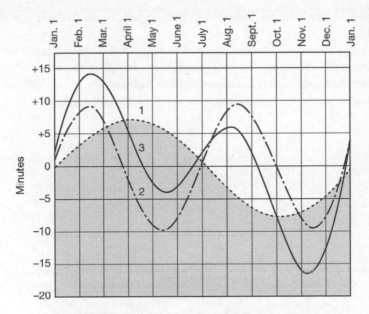

Figure 8: The 'equation of time' curve (3) is the sum of two components. Because the Earth's orbit is an ellipse, and the Earth–Sun distance waxes and wanes through the seasons, a discrepancy is set up (curve 1). Because the Sun's apparent annual motion through the sky is along the ecliptic, not the equator, it seems to run 'fast' or 'slow' at different seasons of the year (curve 2). Add these together and we get curve 3 – the amount by which, at any date, the apparent or sundial day differs from the mean or clock day.

construct chronometers that could stand up to conditions at sea and determine the longitude within sixty, forty and thirty miles, respectively. Harrison's instrument was tested on a voyage to Jamaica, and on its return to Portsmouth in 1762 it was found to have lost just under two minutes. This meant that the longitude of a ship at sea could be determined within eighteen miles. Harrison therefore claimed the full reward of £20,000, but with its customary carefulness in such matters, the government merely advanced him various sums on account and it was not until 1773, when he was eighty, that he was at last paid in full. He died three years later.

The successful transportation of Greenwich time to any place on the globe eventually led to an important change in the method of time-keeping itself. For, since solar time, based on the rotation of the Earth, varies by four minutes for each degree of longitude, it was

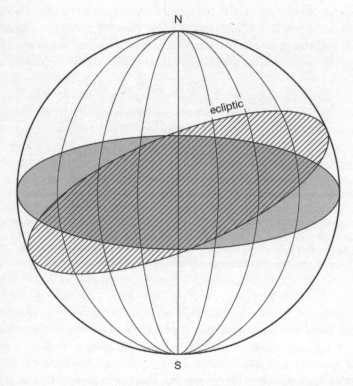

Figure 9: Projection on to the sky of the Earth's equator and meridians of longitude is the conventional way of locating any object in the sky. On this projection the Sun's apparent path, the ecliptic, is a circle cutting the 'equator' twice, at an angle of 23½°, which is the angle of tilt of the Earth's axis relative to the plane of its orbit.

found necessary by 1885 to divide the globe into a series of standard time-zones. With the advent of rapid air transport, this has led in recent years to an entirely new kind of illness, known as *flight dysrhythmia*, or time-zone fatigue, suffered by many of us when we make a long journey in an easterly or westerly direction. This complaint is a consequence of the resulting discord between external local time and the body's metabolic clock that regulates the ups and downs of energy output, digestion and other functions. When a traveller leaves New York at noon and finds himself seven hours later in London, where it is midnight, the metabolic clock begins ringing food and energy alarms – at the wrong time. The result is a feeling of tiredness and irritation that lasts a day or two before the body adjusts to it.

Although in everyday life it is convenient to divide the globe into different time-zones, for astronomical and geophysical purposes scientists throughout the world use the same time, known as *Universal Time* (UT). It is defined as the mean solar time of the Greenwich meridian, and is reckoned on a twenty-four hour basis starting at midnight. Another device that has been found useful for chronological reckoning over long periods of time is to count in *Julian days*, as first proposed by the great classical scholar J. J. Scaliger in 1582. Each Julian day begins at 12 hours UT, starting from day 0 on 1 January 4713 BC. The Julian day that began at this time on 1 January 1970 was numbered 2,440,588. The advantage of using Julian days is that we thereby avoid the irregularities in the lengths of the months and the years.

Precise time

Before the Royal Observatory was founded it was generally assumed by astronomers that the Earth's diurnal rotation is uniform. The first Astronomer Royal, John Flamsteed, realized that this might not be so. As early as 1675, in a letter to Richard Towneley, he wrote 'it is questionable whether the daily return of any meridian on our earth to a fixed star be equal and isochronical at all times of the year'. In fact, although civil time (as distinct from astronomical time) is still based on the rotation of the Earth, we now know that this is subject to small irregularities. The Earth is a solid body surrounded by water and air, and its time of rotation varies slightly from one season to another. It is also being slowed down very gradually by the frictional effect of the tides. In addition, other small changes occur from time to time in the Earth's rate of rotation that are unpredictable.

In 1963 J. W. Wells of Cornell University pointed out that the annual growth bands on fossil corals are made up of daily growth ridges. From counts of the number of daily increments per annual band the number of days in a year has been determined back to the middle of the Devonian period nearly four hundred million years ago. It has been further deduced that the length of the day was less than twenty-one hours six hundred million years ago.

Besides rotating on its axis, the Earth revolves about the Sun. The natural unit of time provided for man's use by this motion is called the *tropical year*, and is defined as the time between two successive passages of the Sun through the spring equinox. This is the point on the celestial sphere where the Sun crosses the projection on to this sphere of the Earth's equator in the spring. It is on the year defined in this way that the seasons and the calendar depend, but it is not the same as the time between successive passages by the Sun through the same fixed point in the sky because the equinox has a retrograde motion of 50.2 seconds of arc a year. This precession of the equinoxes, as it is called, is due to the gravitational pull of the Sun and Moon on the Earth's equatorial bulge, causing the Earth's axis to precess like the axis of a spinning top, with a period of about 25,800 years. The tropical year is equal to 365.2422 mean solar days, whereas the sidereal year is equal to 365.2564 of these days.

Because of the unpredictable small variations in the Earth's rate of rotation, astronomers decided to introduce in 1956 a more accurate unit of time based on the Earth's motion around the Sun.

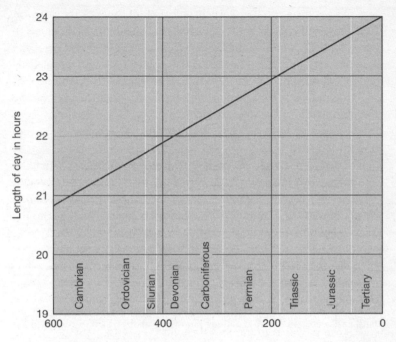

Figure 10: The gradual slowing of the Earth's rotation results in a lengthening of the day from less than twenty-one hours in the Cambrian era to twenty-four hours today.

This unit, known as the second of *ephemeris time*, differs slightly from the average value of the mean solar second. It is defined as the 31556925.9747 part of the tropical year 1900.

In the last few years, because of the growing demand for high-precision measurement, it has become desirable to have some more fundamental standard of time than any that can be derived from astronomical observations. One such standard is the natural period of characteristic electromagnetic waves produced by a vibrating atom or molecule. These electromagnetic waves, due to particular modes of vibration, are of very precise frequency and form sharp 'lines' in the spectrum. Optical spectral lines are unsuitable for use as

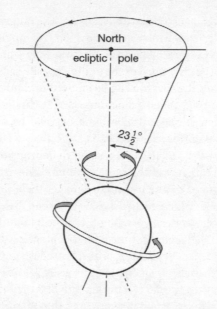

Figure 11: The axis of the Earth, like that of a spinning top, traces out a circular path as it precesses. There is nothing immutable about which star is called the Pole star.

a standard of time because we have no means of measuring their frequencies directly. Certain atoms, however, produce radio frequencies and these can be measured directly. More than ten years ago, this discovery led the British physicist Dr L. Essen to develop the caesium atomic clock or frequency standard. Atoms of caesium produce radio waves of about 9,200 megacycles a second, corresponding to a wavelength of about 3 centimetres. This lies conveniently within the range of wavelengths used for radar, and electronic techniques for handling these frequencies have been highly developed.

The particular type of vibration used in the caesium clock is very different from that of the swinging pendulum. A caesium-133 atom has a fairly heavy nucleus surrounded by a number of shells, like the skins of an onion, each filled with electrons. The outermost shell has only a single electron, which spins like a top. The nucleus also spins and there are two possibilities: the electron may spin in the same sense (direction) as the nucleus, or in the opposite sense. By supplying a little energy, the electron can be induced to change the sense in which it is spinning. When, some time later, it changes back again, the energy given is released as a burst of radio waves with a frequency of 9,200 megacycles a second.

The clock consists, in effect, of a small radio transmitter tuned in to match the frequency of the caesium atoms. The oscillating magnetic field produced by the transmitter makes these atoms resonate when the frequency is correct, just as a singer can make a wine glass vibrate and even shatter by singing into it at its natural frequency. The magnetic field can be tuned to the atomic vibrations with extreme sharpness. The general principle underlying the accurate measurement of radio frequencies can be likened to that used in tuning a piano with the aid of a tuning fork and the human ear. The role of the ear is taken by a radio valve and that of the tuning fork by a standard of frequency that is kept continually running. From this is built up another frequency that is close to the one that has to be determined, the small difference giving rise to a slow beat that is easily measured. For example, if the resulting beat associated with the measurement of a frequency of 200,000 cycles a second were 2 cycles and this were measured to 1 per cent with the aid of only a stopwatch, the over-all accuracy for this very simple measurement would be one part in ten million. From this example we see why it is that the measurement of frequency can lead to the most precise of all physical measurements. Moreover, a standard of frequency has the great advantage, compared with other standards such as that of length, that it can be communicated by radio transmission and so can be made readily available at any place where there is a suitable receiver.

In the case of the caesium clock, the timing can be made to an accuracy of five parts in ten million million, which corresponds to a clock error of only one second in 150,000 years. This is wholly

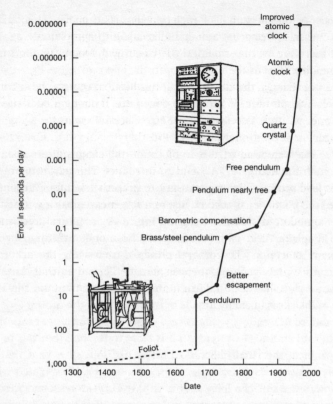

Figure 12: From the Dover Castle clock of the mid-fourteenth century, which could lose or gain as much as a quarter of an hour a day, to the ultimate sophistication of the atomic clock, theoretically achievable time-keeping accuracy has improved at a roughly exponential rate.

independent of astronomical determinations of time and is reproducible with an accuracy which is better than any that has been obtained from such determinations. Consequently in 1967, a new definition of the second was introduced in terms of the natural period of the atom rather than in terms of the motion of heavenly bodies. It was in fact defined as the duration of 9,192,631,770 periods of the caesium-133 radiation discussed above. This figure was chosen to bring the atomic time scale into approximate agreement with the average rate of GMT (Greenwich Mean Time) during the nineteenth century.

During the past decade the Earth has been running slow in comparison with its average rate by amounts which have varied between 1.8 and 3.2 milliseconds a day. Some users of radio time signals require Earth rotational time (GMT), others require a uniform time (TAI). The system adopted is to maintain the signals at a constant rate (corresponding to TAI) and to introduce step adjustments in time (leap seconds) when necessary to keep the signals sufficiently close to GMT to satisfy the navigators. For users requiring a better approximation to GMT the difference is given by code to an accuracy of 0.1 second. There is, however, no question of atomic time superseding GMT, which will remain the basis of the world's time as it has for nearly a century. What has been agreed internationally is that, by suitably relabelling the uniform atomic seconds of the time signals, they will be kept in step with GMT.

Heat loss and radioactivity

Although the modern idea of time is based on the notion of linear advancement all the clocks and other systems for the measurement of time discussed so far have depended on essentially cyclic processes. The earliest proposal to use a linear process for the measurement of time seems to have been made by the astronomer Edmund Halley in 1715. He pointed out that the sea had become salt owing to the accumulation of saline material swept down by the rivers, and he regretted that the ancient Greeks had not 'delivered down to us the degree of saltness of the sea, as it was about two thousand years ago', so that the difference between the saltness then and now could be used to estimate the age of the oceans.

Halley's suggestion was revived towards the end of the nineteenth century. Assuming that the primeval ocean waters were fresh and that the present annual amount of dissolved sodium removed by rivers from the land could be taken as the average throughout geological time, John Joly calculated in 1899 that about 90 million years had elapsed since the oceans were formed. This estimate we now realize was far too low, partly because sodium liberated from the rocks does not all accumulate in the sea but some is recycled by evaporation, or by being blown inland, and some is deposited in marine sediments. Also it is most unlikely that the rate of increase of salinity of the oceans has been uniform. At present the land areas of the world are thought to be much more elevated than in the past, and rivers are much more active in consequence. The present rate at which sodium is being deposited by them into the sea is therefore probably considerably greater than the average in the past.

The sodium clock suggested by Halley is ineffective because of our inadequate knowledge of the factors involved in its operation. Other non-cyclic clocks proposed in the nineteenth century for the measurement of the ages of the Earth and the Sun were based on assumptions that appeared to be difficult to reject. These measurements concerned heat loss. Temperatures in deep mines revealed a fairly uniform increase with depth, indicating that heat flows from the Earth's hot interior to the cooler outer crust where it escapes. This heat loss can be measured, and Lord Kelvin argued that the Earth must be cooling and therefore was hotter in the past. From the present rate of heat flow he calculated the order of magnitude of the time that has elapsed since the Earth's surface was molten and concluded that it was between 20 and 40 million years.

This result was in general agreement with an independent calculation of the age of the Sun. Following the establishment of the law of the conservation of energy, Helmholtz had considered the problem of the source of the Sun's radiation. Chemical burning was quickly ruled out as utterly inadequate. Helmholtz suggested that the only mechanism that could maintain the Sun's heat for more than a few thousand years was the release of energy through gravitational contraction. On this hypothesis, Kelvin calculated that the shrinkage of the Sun to its present size could not have allowed it to radiate for more than about 50 million years.

Kelvin's calculations caused dismay to both geologists and the followers of Darwin, whose ideas necessitated temperature conditions suitable for the maintenance of life on the Earth's surface for far longer periods of time. Nevertheless, these calculations could not be refuted until new sources were found for the maintenance of the Earth's and the Sun's heat. The discovery of radioactivity by Henri Becquerel in 1896, and the subsequent investigation of its role in geological processes, resolved the dilemma. Lord Rayleigh calculated the heat generated by radioactive minerals in the Earth's crust and showed that it easily accounted for the flow of heat at the surface. It was therefore clear that previous estimates of the Earth's age based on terrestrial cooling were far too short. The line of research inaugurated by Becquerel's discovery also led eventually to an understanding of the thermonuclear processes now believed to be responsible for the Sun's heat. These imply a far longer time-scale than that envisaged by Helmholtz and Kelvin.

The discovery of radioactivity was literally epoch-making since it led to new and far more precise methods for the measurement of geological time. In 1902 Rutherford and Soddy announced their famous law that the number of atoms of a radioactive element which disintegrate in unit time is proportional to the number of atoms of the element that are present. This law is based on the fact that the probability of an atom breaking up is independent of its age and also of the temperature, pressure and other physical characteristics of its environment. It depends only on the particular element concerned, because radioactivity is due to instabilities in the structure of nuclei and its rates are determined solely by these instabilities. In other words, radioactive decay is essentially a nuclear phenomenon that involves energies far larger than those corresponding to the chemical bonds and the various external physical influences to which the element might be subjected. It follows that the rate of decay of the nucleus of a given element can be used as a means of measuring time. It is the outstanding example in nature of a non-cyclic linear clock.

In practice two kinds of nuclear clock are used for time-measurement: the *decay clock* and the *accumulation clock*. Except for uranium, actino-uranium and thorium, all naturally occurring radioactive elements are produced from other elements by some nuclear reaction. When a radioactive element is being produced at a constant rate in a given environment it increases in amount until there is a state of secular equilibrium, or balance, between produc-tion and decay, so that the total amount present is constant. If a part is then removed and stored so that no more material is being added to it, this part will decay according to the Rutherford–Soddy law, and the ratio of the amount left after a given time to the amount origin-ally isolated will be a measure of this time.

A convenient unit of measurement is the *half-life*, defined as the time required for one half of any given amount of the element to decay. In virtue of the Rutherford–Soddy law this is a fixed interval of time for a particular element, being in principle independent of the amount present. In practice, decay clocks often do not function satisfactorily after more than about ten half-lives, because of the probable error in determining the amount of material still left (about one part in a thousand of the original amount).

The best known example of a decay clock is the carbon-14 clock. This element is produced in the upper atmosphere as the result of its continual bombardment by cosmic ray particles which produces much nuclear debris, including some neutrons. These neutrons are absorbed by the nitrogen-14 in the atmosphere which then emits a proton and thus changes into carbon-14. This radioactive isotope of ordinary carbon has a half-life of about 5,700 years, which is so short that it is safe to assume that no carbon-14 is primordial. Newly produced carbon-14 is rapidly incorporated into the carbon dioxide of the atmosphere and thus assimilated into the carbon cycle, so that a plant or any other organism that absorbs carbon dioxide receives a proportional share of this radioactive carbon. When the organism ceases to absorb carbon dioxide – for example, when a plant dies –

the carbon-14 clock effectively begins to tick, because the proportion of radioactive carbon to ordinary carbon steadily diminishes according to the decay law. This method of dating depends on two important assumptions: that the rate of production of carbon-14 in the atmosphere has remained sensibly constant over the period to which the method can be applied (up to about forty or fifty thousand years), and that the assimilation of carbon-14 by the organism was rapid compared with the time to be measured. On the whole, these assumptions are usually satisfied, although it has been found in recent years that the former has to be modified somewhat because of fluctuations that have occurred in the intensity of the cosmic ray influx from outer space. Also account must be taken of man's interference with the amount of carbon in the atmosphere, by the burning of coal and petroleum and the explosion of thermonuclear devices. Despite these complications, radio-carbon dating has provided archaeologists and others with a powerful new tool which has yielded results of fundamental importance. For example, it has enabled the wrappings of the Dead Sea scrolls to be dated with an accuracy of ± 100 years, and has led to the surprising discovery that man first appeared in North America only about 11,000 years ago and entered Central and South America soon afterwards.

The other type of nuclear clock depends on the accumulation of some daughter element produced by the decay of a radioactive parent element. The clock depends on the assumption that the number of atoms of the daughter element present at a given epoch is equal to the number of atoms of the parent element that have disintegrated, the latter number yielding a time that can easily be calculated by means of the Rutherford–Soddy law. The assumption is valid provided that there has been no interference with the system from outside, and also that no atoms of the daughter element were already present when the system was formed. Of these limitations, the second is usually the more serious. Many rocks are believed to satisfy the first condition well enough for the purposes of age determination, but some atoms of the same element as those produced by the decay process were usually present initially. Consequently, instead of assuming that there were none then, methods must be devised for estimating how many in fact there were. The measured ratio of the amounts of parent and daughter elements present at a given time can thus be used to determine the age of the system.

Of the many long-lived radioactive elements only uranium, potassium and rubidium have given rise to accumulation clocks that have been much used. The ultimate stable end-products of uranium decay are helium and lead. As early as 1906 Rutherford made the first attempt to measure the ages of minerals from their uranium–helium ratio. He obtained an age of 500 million years for two samples of uranium-bearing minerals, but he realized that this must be a minimum limit for their ages because these minerals are not compact but porous, and some of the helium must have escaped. It has since been abundantly confirmed that most uranium–helium ages are too low and the method is now seldom used.

In 1907 B. B. Boltwood, an American chemist, published uranium–lead ages for several minerals ranging from 410 to 2,200 million years. Although later knowledge revealed that his ignorance of various complicating factors led to results that were usually somewhat too high, his work made it clear that the uranium–lead accumulation clock could be used to construct a reasonably reliable quantitative scale of geological time. The successful use of this accumulation clock is due to the very long half-life – 4,500 million years – of uranium-238. In recent years our knowledge of geological time obtained in this way has been confirmed and supplemented by the use of potassium–argon and rubidium–strontium clocks.

Scales of time

Implicit both in the development of modern ultra-precise artificial clocks (such as the caesium clock), which are suitable for the accurate measurement of very short intervals, and in the use of naturally occurring radioactive clocks, for the determination of very long intervals, is the assumption that all atoms of a given element behave in exactly the same way, irrespective of place and epoch. The ultimate scale of time is therefore based on our concept of universal laws of nature. This was already recognized in principle before these practical developments occurred, notably by the great French mathematician and philosopher of science Henri Poincaré (a cousin of the French statesman and President of France in the First World War, Raymond Poincaré).

In one of his essays, written nearly seventy years ago, he argued that, although we have a perfectly clear idea of what we mean when we say that one conscious fact is earlier than, later than, or simultaneous with another, we have no direct intuition of the equality of two intervals of time. Instead, any assertion of the equality of two intervals of time depends on a definition that involves a certain degree of arbitrary choice. This is because even the best physical chronometers must be corrected every now and then. He pointed out that astronomers were not content with the rotation of the Earth as an exact measure of time, since the tides tend to slow it down. Because of this there is an apparent slight speeding-up of the motion of the Moon. This speeding-up was calculated on the basis of Newton's laws. In other words, it was assumed that time should be so defined that Newton's laws are verified. If some other way of measuring time were adopted, Newton's laws would probably assume a more complicated form. Consequently, argued Poincaré, the definition implicitly adopted by the astronomers meant that time should be so defined that the equations of mechanics are as simple as possible. To quote his own words, 'there is not one way of measuring time more true than another; that which is generally adopted is only more *convenient*. Of two watches we have no right to say that one goes true, the other wrong; we can only say that it is advantageous to conform to the indications of the first.'

Poincaré appears, however, to have overlooked the possibility that different physical laws may in fact entail different scales of time that do not keep pace with each other. For example, is the time measured by the law of radioactive decay of uranium-238 with a half-life of about 4,500 million years the same as that implied by Newton's laws of dynamics and gravitation which have been used to study the motions of the planets only during the last 300 years? This is a question to which we do not yet know the answer.

5

Time and relativity

THE INVENTION OF MECHANICAL CLOCKS which could, if properly regulated, tick away continually for years on end greatly influenced belief in the uniformity and continuity of time. These characteristics were implicit in the idea of physical time adopted by Galileo in the dynamical part of his famous *Discourses on Two New Sciences*, published in 1638. Although he was not the first to represent time by a geometrical straight line, he became the most influential pioneer of this idea through his theory of motion expounded in this book.

For the first explicit discussion of the concept of mathematical time we must go to the *Geometrical Lectures* of Isaac Barrow, written about thirty years after the publication of Galileo's book. His views on the nature of time are not only of great interest in themselves, but are important because of their influence on Newton, who succeeded him in the Lucasian chair of mathematics at Cambridge in 1669. Barrow maintained that, because mathematicians frequently make use of time, they ought to have a clear idea of the meaning of the word, for otherwise, he said, they are quacks! Although time is measurable by motion, he was careful to distinguish the two:

> Time denotes not an actual existence but a certain capacity or possibility for a continuity of existence; just as space denotes a capacity for intervening length. Time does not imply motion, as far as its absolute and intrinsic nature is concerned; not any more than it implies rest; whether things

move or are still, whether we sleep or wake, Time pursues the even tenour of its way. Time implies motion to be measurable; without motion we do not perceive the passage of Time. We evidently must regard Time as passing with a steady flow; therefore, it must be compared with some handy steady motion, such as the motion of the stars and especially of the Sun and Moon.

Barrow pointed out that

strictly speaking the celestial bodies are not the first and original measures of Time; but rather those motions, which are observed round about us by the senses and which underlie our experiments, since we judge the regularity of the celestial motions by the help of these. Not even is Sol himself a worthy judge of Time, or to be accepted as a veracious witness, except so far as time-measuring instruments attest his veracity by their votes.

Barrow regarded time as essentially a mathematical concept that has many analogies with a line, for it has length alone, is similar in all its parts and can be regarded either as a simple addition of successive instants or as the continuous flow of one instant. He thought it could be represented by, to quote his own words, 'either a straight or a circular line'. Although the reference here to 'a circular line' shows that Barrow was not completely emancipated from traditional ideas, his statement goes further than any of Galileo's, for Galileo used only straight-line segments to denote particular intervals of time. Barrow took care, however, not to push the analogy between time and a line too far. Time, in his view, was 'the continuance of anything in its own being', and in a passage to which reference will be made later, he remarked, 'nor do I believe there is anyone but allows that those things existed equal times which rose and perished together'.

Barrow's idea that, irrespective of 'whether things move or are still, whether we sleep or wake, Time pursues the even tenour of its way' is echoed in the famous definition at the beginning of Newton's

Principia of 1687. 'Absolute, true and mathematical time', wrote Newton, 'of itself and from its own nature, flows equably without relation to anything external.' Newton admitted that in practice there may be no such thing as a uniform motion by which time may be accurately measured, but he thought it necessary that, in principle, there should exist an ideal rate-measurer of time. Consequently, he regarded the moments of absolute time as forming a continuous sequence like the points on a geometrical line, and he believed that the rate at which these moments succeed each other is independent of all particular events and processes.

Newton's idea of absolute time existing in its own right accords with the commonsense idea that most of us nowadays accept automatically when we try to think about time. We feel that time is something that can have neither beginning nor end and must continue independently of whatever happens. Newton's views made a great impression on the philosopher John Locke, whose *Essay Concerning Human Understanding* was published in 1690, only three years after Newton's *Principia*. In it we find the clearest statement of the scientific conception of time that was evolved in the seventeenth century. Locke wrote:

> Duration is but as it were the length of one straight line extended *in infinitum*, not capable of multiplicity, variation or figure, but is one common measure of all existence whatever, wherein all things, whilst they exist, equally partake. For this present moment is common to all things that are now in being, and equally comprehends that part of their existence as much as if they were all but one single being; and we may truly say, they all exist in the same moment of time.

Despite its appeal to laymen, Newton's idea of absolute time as flowing at a uniform rate independently of all that actually goes on in the world – so that it would continue just the same even if the universe were completely empty – has been frequently and justly criticized by philosophers. It assumes that time is a kind of thing and ascribes to it the function of flowing. If time were something that flowed, then it would itself consist of a series of events in time, but this would be meaningless. Moreover, if time can be considered in isolation 'without relation to anything external', as Newton says, what meaning could be attached to saying that its flow is not uniform, and if no meaning can be attached even to the possibility of non-uniform flow, what point is there in saying that it 'flows equably'?

It may be objected that this criticism is all very well but that it misses the point that Newton was not a philosopher in the modern professional sense of the term but was a scientist who was primarily concerned with the practical use of his fundamental ideas. Unfortunately, however, his definition of absolute time is of no practical use! We can only observe events and actual processes in nature and base our measurement of time on them. Newton's concept of time implies that there exists a unique series of moments and that events are distinct from them but can occupy some of them. Newton was driven to accept this concept, not merely because of his desire for an ideal measure of time to compensate for the difficulty of determining a truly accurate *practical* time-scale, but because he was convinced that there must be an ultimate absolute time in nature.

The idea that moments of absolute time exist in their own right was rejected by Newton's contemporary Leibniz, who argued instead that events are more fundamental. In his view, moments are merely abstract concepts, being classes or sets of simultaneous events. He defined time not as a thing in itself but simply as the order in which events happen. Leibniz based his philosophy of time on the principle that nothing happens without there being a reason why it should be so rather than otherwise. He argued as follows:

> Suppose someone asks why did not God create everything a year sooner and that he wants to infer from this that God has done something for which he could have had no reason for doing it when he did rather than at some other time. This inference would be correct if time existed independently of things. For then there would be no reason why things should

exist at certain instants and not at others, their succession remaining the same.

Leibniz claimed that this shows the absurdity of imagining that instants can exist when there are no things. Consequently, there can be no way of distinguishing between the universe as it actually is and as it would be if it had been created a year sooner.

Leibniz's theory that events are more fundamental than moments is known as the *relational theory of time*. It is based on the idea that we derive time from events and not the other way round. This means, for example, that we should regard two events as simultaneous not because they occupy the same moment of absolute time but because each occurs when the other does. For the temporal correlation of events that are not simultaneous we can use the following idea: we can regard all simultaneous events as forming a particular state of the universe, and these states as occurring one after the other like yesterday, today and tomorrow. Leibniz's theory is nowadays regarded as more acceptable than Newton's because, as we shall see, it is more in accord with modern developments in physics.

In the eighteenth and nineteenth centuries, however, the Newtonian point of view was dominant, so that by the beginning of the present century it had come to be generally assumed that there is but one universal system of time and that it exists in its own right. This belief was not confined to scientists but was fostered by the growing tendency in industrial civilization for men's lives to be regulated by the clock, particularly following the mass production of cheap watches. Even the division of the Earth's surface into separate time-zones did little to undermine belief in the absolute and universal nature of time. The introduction of 'daylight saving' (Summer Time) in the United Kingdom in 1916, during the First World War, was met with a storm of protest not only from those who found it a nuisance but also from those who thought it outrageous to interfere with 'God's Own Time'! More sophisticated people realized that the choice of time zero and of time units could be altered to suit man's convenience, but they believed that these were the only arbitrary features in our conception of time and that everything else about it was unique and unalterable. Time was in fact regarded as a kind of moving knife-edge covering all places in the universe simultaneously. There was general agreement with the views of John Locke, quoted above, that 'duration is one common measure of all existence whatever', and that 'the present moment is common to all things now in being'. It therefore came as a great shock when, in 1905, Einstein discovered a previously unsuspected gap in the theory of time measurement which made him reject these assumptions and the whole philosophy of time associated with them.

Clocks at rest and clocks in motion

The starting point of Einstein's investigations into the nature of time was his desire to reconcile James Clerk Maxwell's electromagnetic theory of light with the rest of physics based on Newton's laws of mechanics. In one of the corollaries to these laws as set out in his *Principia* of 1687, Newton states that 'The motions of bodies included in a given space are the same among themselves, whether that space is at rest or moves uniformly forward in a straight line.' This means that all purely mechanical experiments must give the same results whether they are performed in a laboratory at rest on the Earth's surface or on, say, a ship moving steadily in a given direction. For example, a stone let fall from one's hand descends in a straight line with the same constant acceleration in both cases. Although this principle of relativity was universally recognized to be valid for material bodies, it seemed to be at variance with Maxwell's theory of electromagnetic radiation. Instead, light and other electromagnetic effects were regarded as wave-like disturbances propagated through a stationary universal medium, the luminiferous ether, with a velocity, usually denoted by the letter c, of about 300,000 kilometres a second. Nevertheless, despite the great success of Maxwell's theory – in particular, it led Hertz to demonstrate the existence of radio waves – there were some serious conceptual difficulties associated with the idea of the luminiferous ether.

One of the most puzzling occurred to Einstein when he was only sixteen. He tried to imagine what he would observe if he were to travel through the ether with the same velocity as a beam of light. According to the usual idea of relative motion, the beam of light should then appear as a spatially oscillating electromagnetic field *at rest*. But such a concept was unknown to physics and at variance with Maxwell's theory. Einstein began to suspect that not only the laws of mechanics but all the other laws of physics, including those concerning the propagation of light, must remain the same for all observers however fast they move, even if this conflicts with the hypothesis of the luminiferous ether. He thus began to feel convinced that the principle of relativity applies to electromagnetic as well as to mechanical phenomena and that the velocity of light is not only the same for all observers at relative rest but also for those in uniform relative motion.

Nevertheless, it was only after years of thought that Einstein finally felt compelled to accept this conclusion, for it conflicted with traditional ideas concerning the measurement of motion. Eventually he began to analyse the assumptions underlying the way in which we measure motion and he saw that it must depend on how we measure time. It occurred to him that time measurement depends on the idea of simultaneity. To quote his own words, 'all judgments in which time plays a part are always judgments of *simultaneous events*'. Although this statement is too sweeping as it stands, what Einstein meant was that all *measurements of duration* involve judgements of simultaneity – that is, of the coincidence in time of an event with, say, a particular position of the hands of a watch. Suddenly he was struck by the fact that, although this idea is perfectly clear without further discussion when we are concerned with events in our immediate vicinity, it is not so for distant events. The crucial passage in his 1905 paper runs as follows:

If we wish to describe the motion of a material point, we give the values of its coordinates as functions of the time. Now we must bear carefully in mind that a description of this kind has no physical meaning unless we are quite clear as to what we understand by 'time'. We have to take into account that all judgments in which time plays a part are always judgments of *simultaneous events*. If, for instance, I say 'that train arrives here at seven o'clock', I mean something like this: 'the pointing of the small hand of my watch to seven and the arrival of the train are simultaneous events.'

It might appear possible to overcome all the difficulties attending the definition of 'time' by substituting 'the position of the small hand of my watch' for 'time'. And in fact such a definition is satisfactory when we are concerned with defining a time exclusively for the place where the watch is located; but it is no longer satisfactory when we have to connect in time series of events occurring at different places, or – what comes to the same thing – to evaluate the times of events occurring at places remote from the watch.

Einstein realized that the concept of simultaneity for a distant event and one close to the observer depends on the relative position of the distant event and the mode of connection between it and the observer's perception of it. If the distance of an external event is known and also the velocity of the signal that connects it and the observer, he can calculate the epoch at which the event occurred and can correlate this with some previous instant in his own experience. This calculation will be a distinct operation for each observer, but until Einstein raised the question it had been tacitly assumed that, when we have found the rules according to which the time of perception is determined by the time of the event, all perceived events could be brought into a single objective time-sequence, the same for all observers. Einstein not only realized that it was a hypothesis to assume that, if they calculate correctly, all observers must assign the same time to a given event, but he produced cogent reasons why, in general, this hypothesis should be rejected.

Einstein assumed that there are no instantaneous connections between external events and the observer. The classical theory of time, with its assumption of world-wide simultaneity for all observers, in effect presupposed that there were such connections. Instead, Einstein postulated that the most rapid form of communication is by means of electromagnetic signals – including light rays and radio waves – in empty space, and that their speed is the same for all observers at rest or moving uniformly in straight lines. He regarded this assumption as a consequence of his principle that the laws of physics are the same for all such observers. He found that observers in uniform relative motion would, in general, be led to assign different times to the same event, and that a moving clock would appear to run slow compared with an identical clock at rest with respect to the observer. For the velocities that we encounter in everyday life this effect is negligible, but the nearer the relative velocity of the moving clock to the speed of light the slower it appears to run compared with the clock carried by the observer.

Einstein also found that, according to his theory, Newton's laws of motion, formerly regarded as the foundation of a great part of physics, have to be modified, particularly for rapidly moving bodies. For example, the inertial mass of a body, formerly supposed to be independent of its motion, was now found to increase indefinitely the nearer its velocity approaches that of light. Consequently, a given force acting on the body produces smaller and smaller changes in its velocity the faster it moves, and as a result no particle of matter can ever attain the speed of light. In this way Einstein solved his original problem concerning the observer who moves with the same velocity as a beam of light. This motion is physically impossible. In particular, no clock can be transported in this way: if in fact a clock were to move with this velocity its action would be arrested and it would therefore always record the same time!

It is well known that Einstein's Theory of Special Relativity, as it came to be called, automatically explained the failure of the Michelson–Morley experiment, first performed in 1887, nearly twenty years before Einstein's paper appeared. The object of this experiment by these two American physicists was to determine the motion of the Earth with respect to the ether. Michelson believed that a very sensitive instrument which he had invented, known as the interferometer, would permit him to measure this motion. With this instrument interference patterns were produced when two beams of light were reunited after being transmitted to and fro along arms of equal length, at right angles to each other. These patterns were compared for different orientations of the arms and should have shown some variation depending on the change in these directions relative to the Earth's motion. Despite the fact that in the course of half a year the Earth changes its velocity relative to the Sun from thirty kilometres a second in one direction to the same speed in the opposite direction, whenever the experiment was performed it gave a null result, although the interferometer was so sensitive that it could have detected effects due to a velocity of less than ten kilometres a second. Indeed, if it had been possible to make the experiment in the days when the Copernican theory was still in dispute, it would have been regarded as conclusive evidence that the Earth does not move at all, but stays at rest at the centre of the universe! Of course, this interpretation was quite impossible to sustain by the nineteenth century and some other explanation for the failure of Michelson's experiment had to be sought.

In the early 1890s Fitzgerald in Ireland and Lorentz in Holland independently suggested that Michelson's null result could be explained if it were assumed that the length of a moving body was automatically diminished in virtue of its motion by a certain factor depending on its velocity. This factor, now known as the Fitzgerald–Lorentz contraction, would not be detectable by an observer moving with the body, since all his instruments would be affected in the same way. This point of view led Lorentz to consider the effect of electric forces on the electronic and atomic constitution of matter, with the object of explaining why this contraction must occur in the same way in all forms of matter.

Instead of this complicated attempt at an explanation, it was one of the great merits of Einstein's theory that it *automatically*

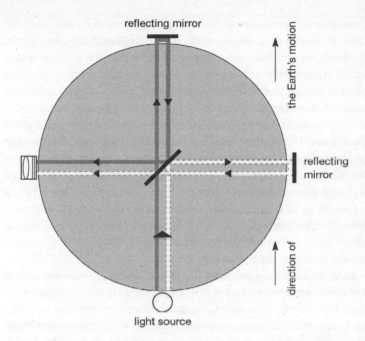

Figure 13: Simplified diagram of the Michelson–Morley experiment. A beam of light is split by a half-silvered mirror into two beams which travel paths of identical length at right angles. Reunited at the interferometer (left), they were expected to form varying interference patterns both as the orientation of the apparatus was changed and as the Earth's direction of travel through the hypothetical 'ether' changed.

accounted for the Michelson–Morley result, since according to his theory the velocity of light is invariant and therefore must be the same along either arm, irrespective of the orientation and motion of the apparatus. Michelson's null result was therefore strong confirmatory evidence for Einstein's theory. Moreover, instead of assuming that there are real structural changes in the constitution of matter due to motion, Einstein's theory only involved apparent changes, that is changes relative to the observer.

The same conclusion also applied to the apparent slowing down of a clock in uniform relative motion as compared with the so-called *proper time* kept by a similar clock at rest relative to the observer. In Einstein's theory, this phenomenon of 'time dilatation', as it is now called, is essentially a phenomenon of measurement, applicable to all forms of matter including living organisms, and is a reciprocal effect in the following sense: if A and B are two observers in uniform relative motion, B's clock seems to A to run slow and equally A's clock seems to run slow according to B. This reciprocal relationship no longer holds if forces are applied to change the motion of one of the observers. In particular, if A and B are together at some instant and at a later instant the motion of B is suddenly reversed by the intervention of a force so that eventually he comes back to A with the same speed as he left, the time that elapses between the instant at which B left A and the instant when he returns to A will be shorter according to B's clock than according to A's.

This extraordinary consequence of his theory of time (often called the 'clock paradox') was pointed out by Einstein in his 1905 paper. Six years later he put it in a more graphic form to emphasize the fact that, the nearer the relative speed to the velocity of light, the greater would be the discrepancy in such a case between the two clocks. He said,

> If we placed a living organism in a box . . . one could arrange that the organism, after an arbitrarily lengthy flight, could be returned to its original spot in a scarcely altered condition while corresponding organisms which had remained in their original positions had long since given way to new generations. In the moving organism the lengthy time of the journey was a mere instant, provided the motion took place with approximately the speed of light.

Consequently, although we accept Isaac Barrow's view that 'Time is the continuance of anything in its own being', the special theory of relativity prevents our agreeing with him unconditionally when he went on to say, 'nor do I believe there is anyone but allows that those things existed equal times which rose and perished together.'

Direct evidence for the existence of time dilatation has come from the study of cosmic-ray phenomena. Elementary particles known as mu-mesons, produced by cosmic-ray showers, disintegrate spontaneously, their average proper lifetime (that is, time from production to disintegration according to an observer travelling with them) being about two microseconds (two millionths of a second). These particles are mainly produced at heights of about ten kilometres above the earth's surface. Consequently, those observed in the laboratory on photographic plates must have travelled that distance. But in two microseconds a particle that travelled with the velocity of light would cover less than a kilometre, and according to the theory of relativity all material particles travel with speeds less than that of light. In fact, the velocity of these mu-mesons has been found to be very close to that of light, the corresponding time-dilatation factor being about ten. This is just the amount required to explain why it is that to the observer in the laboratory these particles appear to travel about ten times as far as they could in the absence of this effect.

In recent years time dilatation has been frequently invoked to explain similar phenomena that have been observed in the case of particles moving close to the speed of light in high-energy accelerators.

Relativistic time and relational time

Einstein's theory of special relativity is incompatible with Newton's concept of absolute time, but it can be regarded as a development of Leibniz's theory of relational time. For, although Leibniz himself envisaged a single time-system, the idea that time is derived from events – which is the essence of his theory – is compatible with the existence of a multiplicity of time-systems associated with different observers.

In formulating his concept of absolute time, Newton referred both to the successive order of events in time and to the rate at which they succeed each other. In his view, these are quite distinct: the temporal order, or before-and-after sequence of events, does not determine the

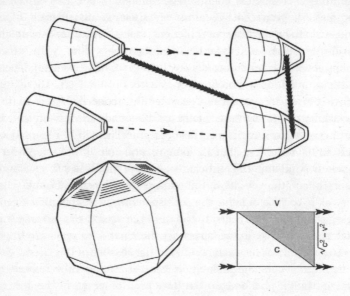

Figure 14: The invariant speed of light, propounded by Einstein, has strange implications for time. Consider two space ships, travelling a parallel course at the same speed, as they pass a space station. At the instant of passing, one space ship sends a flash of light to the other. The receiving ship sees the signal coming straight across, but from the station, which sees the ships moving past, the flash will seem to have crossed diagonally.

We can use Pythagoras' theorem to make this clearer. If the ships are passing the station at v miles a second, in one second they will have moved on v miles. If the velocity of light is c miles a second, the flash will have travelled c miles in that second, as judged from the station. But judged from either ship it will seem to have travelled only along the third side of the triangle, $\sqrt{(c^2 - v^2)}$ miles. This side, being at right angles to the line of motion of the space ships, has the same length for both the observer on the space station and those on either of the space ships. (This is equally true in relativistic and in classical physics.) Measured by either station or ship, the speed of light is the same, which can only mean that, in the ship, time passes more slowly when compared with the space station's clock. And, of course, vice versa, for this is a reciprocal effect.

duration of time that elapses between one event and another. Instead, he believed that the rate at which events succeed each other is determined by the respective moments of absolute time with which they are correlated and what he called the rate of 'flow' of this time. Leibniz's definition of time as the *order* in which events happen makes no mention of the durational aspect of time and therefore, unlike Newton's, is not incompatible with the concept of time dilation. Leibniz's universe was composed of monads ('atoms' endowed, in varying degrees, with powers of perception), which he regarded as mutually independent, but his famous principle of pre-established harmony stipulated that the states of all monads at every instant correspond with each other. Leibniz illustrated this principle by the simile of two clocks that have been so perfectly constructed that they keep perfect time with each other without either mutual influence or external assistance. Consequently, in so far as the temporal aspect of the universe is concerned, Leibniz's principle of harmony is equivalent to the postulate of a single universal time. We must therefore discard this principle if we are to reconcile Leibniz's way of regarding time with Einstein's theory of relativity.

Instead, we stipulate that each observer, at rest or in uniform relative motion, has his own time-system. We can then reformulate, within the framework of the special theory of relativity, Leibniz's theory of time which we have previously discussed only from the classical point of view. First, we define an observer's *proper time* as the order of succession of events that occur locally. In other words, the observer's proper time is the time kept by the clock that he carries with him. Second, to any event that occurs elsewhere he can assign a *coordinate time* calculated from his knowledge of the distance of the event from him in space, the proper time of his perception of it and the speed of transmission of the signal connecting the event in question with him – usually a light signal, radio signal or some other form of electromagnetic signal. In the case of an event that occurs locally the coordinate time coincides with the proper time. For a given observer, all events throughout the universe to which he assigns the same coordinate time define an instantaneous state of the universe; and the order of succession of these states defines time as a whole for him. We thus see that, whereas for Newton time was independent of the universe and for Leibniz it was an aspect of the universe, Einstein's theory leads us to regard it as an aspect of the relationship between the universe and the observer.

We have seen that, according to special relativity, the duration of time between two events depends on the observer, and that this is compatible with Leibniz's idea of time as the order of succession of phenomena if we drop his principle of pre-established harmony. We have therefore reformulated Leibniz's definition of time so that it no longer refers to a single time-system but to the variety of time-systems associated with a plurality of observers. We might still expect that, even if the temporal *interval* between two events E and F depends on the observer, the temporal *order* in which they occur – for example, event E preceding event F – would be independent of the observer, since temporal order is intimately bound up with our idea of causality. On the contrary, however, it is one of the most profound consequences of Einstein's theory that, *in certain circumstances*, temporal order is dependent on the observer.

It is not difficult to describe the circumstances in which this can come about, although a complete account cannot be given without recourse to mathematics (see the Appendix on page 140). Suppose that, according to an observer in whose experience E occurs but F does not, F is later than E and that the time-interval between them is t, their distance apart being r. If r is less than ct, where c is the velocity of light, it is possible for another observer to experience both events locally. All that is required is that the second observer should coincide with the first observer at event E (their clocks being calibrated to read the same time then) and move in the direction of F with a uniform speed given by the ratio of r and t to arrive at the place of F at the instant when F occurs. Since r is less than ct, the relative velocity of this observer will be less than the velocity of light and so is allowed by the theory. Consequently, both events E and F can lie within the local experience of a particular observer, and F occurs after E in his experience, as is shown in the Appendix. The argument can be reversed, so that if F is later than E in the second observer's experience the first observer will attribute to F an epoch that occurs later than that of E. More generally, it is not difficult to prove that for all observers considered in special relativity F will be regarded as occurring later than E.

Similarly, if r is equal to ct, E and F can be connected by a ray of light (or other electromagnetic signal), since the velocity given by the ratio of r and t is exactly c. In this case too F will be regarded as occurring later than E according to all observers. (As we have already remarked, no observer or clock can be transported with the velocity of light.) If, however, r is greater than ct, no observer can be found who experiences both events E and F because his velocity relative to the first observer would have to be greater than c, and this is not allowed. Instead, as is shown in the Appendix, an observer can be found for whom F appears to be simultaneous with E, the time-interval between them being zero. The relative velocity of this observer with respect to the first observer is given by the ratio of c^2t and r, and this is less than c. Moreover, there will be observers, with velocities exceeding this ratio but less than c, for whom the time-interval from E to F will appear to be negative: in other words, F will actually be regarded as occurring *before* E.

The ostensibly paradoxical result that the temporal order of certain events can actually be reversed by appropriate change of observer is closely associated with the fact that in special relativity no causal influence can be transmitted with a speed exceeding that of light, whereas before Einstein formulated this theory it was implicitly assumed that instantaneous transmission could occur. It was therefore essential in pre-relativistic theory that the temporal order of events should be absolute, that is independent of the observer. On the other hand, it is equally essential in relativistic physics that this should be so only for events that can be causally connected by an influence that is transmitted with a speed not exceeding that of light. In this way Einstein's theory of time, despite its apparent paradoxes, is found to be self-consistent, so long as we have no empirical evidence that in nature 'information' can travel faster than light.

6

Time, gravitation and the universe

ALTHOUGH THE RELATIONAL THEORY of time is now generally accepted, as recently as 1901 Bertrand Russell argued against the relational definition of a moment as a particular state of the universe. He claimed that it was not *logically* absurd to imagine the separate occurrence of two identical states of the universe. In fact, as we have seen, this possibility was widely regarded as true before the rise of modern science. Russell argued that if we define a moment as a particular state of the universe we should be faced with the logical absurdity that two moments could be both different and identical.

To avoid this difficulty we are not obliged to fall back on Newton's concept of absolute time. Instead, Russell's argument pinpoints an essential distinction between the ideas of a cyclic *universe* and truly cyclic *time*. The cyclical concepts of time that were so widely accepted in past ages were, strictly speaking, expressions of the idea that the universe keeps repeating the same cycles of processes and events.

If, on the other hand, time itself were truly cyclic, it would be closed like a ring. This idea is nonsense. For if time were cyclic in this sense, there would be no difference between the universe going through a single cycle of events and through a sequence of identical cycles, since any difference would imply that there is a basic non-cyclic time to which the different cycles could be related and distinguished one from another. Moreover, the same argument would also apply to the initial and final events of a single cycle. For, if they were truly identical, there would be no sense in regarding them as occurring separately. Indeed we cannot even make a round trip in ordinary linear time in the way we can in space, for if we could it would mean that we could travel into our own past and do something to ourselves which we already knew, by our own memory, had in fact never happened to us. Consequently, should the universe repeat itself, we must regard the repetition of a given stage in its history as an event distinct from its previous occurrence. This must be so, even if time is relational and not absolute. But it would seem to imply that time must be regarded as a fundamental feature of the universe that is not reducible to anything else: the date becomes an essential characteristic of an event.

Space-time

The modification of the relational theory of time made in response to the demands of Einstein's special theory of relativity introduces a new complication into this argument. For if time is an aspect of the relationship between the observer and the universe, a variety of dates can be assigned to the same event depending on the choice of observer. How then can the date be an essential characteristic of an event? Einstein's critique of the classical concept of simultaneity seems to dispose of the possibility of an *objective* sequence of temporal states of the universe, since each observer has his own sequence of these states and none would appear to be in any way specially privileged.

Despite there being no uniquely privileged observer in the special theory of relativity, there is a privileged *class* of observers, namely all those that are at rest or in uniform relative motion. The motion of these observers is said to be *inertial*, and their frames of reference (hypothetical grids for locating positions relative to those observers) are called *inertial frames of reference*. In classical physics Newton's laws of motion, for instance the law which states that a particle not acted on by an external force is either at rest or in uniform motion, are valid only for observers whose frames of reference are inertial.

Similarly, in the special theory of relativity the new laws of motion that replace Newton's laws are formulated with respect to inertial frames of reference. Normally, in special relativity we consider only inertial frames, but in Einstein's general theory of relativity, which he introduced some ten years after the special theory, all possible frames of reference in all possible kinds of motion are assumed to be on an equal footing. According to this *principle of general covariance*, as it is called, there are no privileged observers whatsoever. Consequently, the difficulty of assigning a unique date to a given event became even more pronounced, and it was therefore not surprising that, although in his earlier work Einstein made the greatest contribution since the seventeenth century to our understanding of time, in his later work this concept came to play a definitely subordinate role.

The essential link in the development of Einstein's ideas about time between the publication of his two theories of relativity was provided by the concept of space-time introduced by the celebrated mathematician Hermann Minkowski and expounded by him in a famous lecture to a scientific gathering in Cologne in 1908. Following Einstein, Minkowski argued that 'nobody has ever noticed a place except at a time, or a time except at a place.' A point of space at a point of time he called a 'world-point', and the totality of all conceivable world-points he called the 'world'. A particle of matter or electricity enduring for an indefinite time will correspond in this representation to a curve which he called a 'world-line', the points of which can be labelled by successive readings of the time that would, at least in principle, be exhibited by a clock carried by the particle. 'The whole universe', he claimed, 'is seen to resolve itself into similar world-lines', and he suggested that 'physical laws might find their most perfect expression as reciprocal relations between these world-lines'.

Minkowski's object was to find a new substitute for the Newtonian absolute time and space discarded by Einstein. In their place he

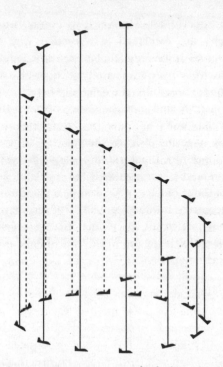

Figure 15: An airliner circling an airport can be depicted either as describing a circle in three-dimensional space, or, with the time dimension added, it can be represented as a helix. This is an instance of a world-line in space-time.

advocated his absolute 'world' which gives different 'projections' in space and time for different observers in uniform motion. This absolute 'world' was later called *space-time*.

The essence of Minkowski's analysis can be briefly described. In pre-relativistic physics the spatial distance and the time interval between two given events occurring at different places at different times are both invariant, that is they have the same values for all observers. In special relativity, as we have seen, neither is an invariant, their values with respect to different observers in uniform relative motion being subject to the Fitzgerald–Lorentz contraction and time-dilatation effects. Minkowski found that, if the spatial distance and temporal interval between any two events (world-points) in special relativity are combined in a certain way, the resulting 'space-time interval' is the same, that is invariant, for all observers in inertial motion. This invariant space-time interval is defined by the rule that its square is equal to the difference between the squares of the temporal interval and spatial distance between the two world-points concerned, units of space and time measurement being chosen so that in terms of them the velocity of light is unity. In Minkowskian space-time the world-line of any observer or particle in uniform motion or at rest is represented by a straight line. One of the peculiarities of the geometry of Minkowski's world is that, whereas in ordinary geometry no line segment is of zero length, there are world-lines along which the space-time distance vanishes. They are known as light-paths, being the world-lines of light and other forms of electromagnetic radiation.

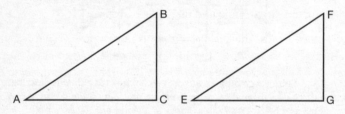

Figure 16: In Euclidean space, if AC and BC are at right angles, then by Pythagoras' theorem $AB^2 = AC^2 + BC^2$. In Minkowskian space-time, if E and F are two events, EG is the time interval between them and GF their spatial separation, then if units of measurement are chosen so that the velocity of light is unity, the space-time interval EF is such that $EF^2 = EG^2 - GF^2$. It will be noticed that if the triangle is isosceles, and $EG = GF$, then EF is zero.

Minkowski's concept of space-time has proved to be one of the most valuable contributions ever made to theoretical physics by a mathematician. In his enthusiasm Minkowski exclaimed, 'Henceforth space by itself, and time by itself, are doomed to fade away into mere shadows, and only a kind of union of the two will preserve an independent reality.' This famous, but excessive, claim tended to reduce the importance of time much more than that of space. Indeed, Minkowski's space-time was envisaged as a new kind of hyper-space in which events do not happen but we merely 'come across them'. As Hermann Weyl expressed it,

> The scene of action of reality is not a three-dimensional space, but rather a *four-dimensional world, in which space and time are linked together indissolubly*. However deep the chasm may be that separates the intuitive nature of space from that of time in our exprerience, nothing of this qualitative difference enters into the objective worlds which physics endeavours to crystallize out of direct experience. It is a four-dimensional continuum which is neither 'time' nor 'space'. Only the consciousness that passes on in one portion of this world experiences the detached piece which comes to meet it and passes behind it as history, that is, as a process that is going forward in time and takes place in space.

In other words, the *passage* of time is merely to be regarded as a feature of consciousness that has no objective counterpart. Weyl's view, like Minkowski's, was essentially that of the 'block universe', to use the term coined by the American psychologist and philosopher William James to denote the hypothesis that the world is like a film strip: the photographs are simply there and are merely being exhibited to us. In this view, even if, as Weyl says, the four-dimensional continuum is neither 'time' nor 'space', the guiding concept is evidently more spatial than temporal.

Following Minkowski's lead, Einstein came to the conclusion that the objective world of physics is essentially a four-dimensional structure, its resolution into three-dimensional space and one-dimensional time not being the same for all observers. 'It appears, therefore, more natural', he wrote, 'to think of physical reality as a four-dimensional existence, instead of, as hitherto, the *evolution* of a three-dimensional existence.'

Einstein and gravitation

The principle of general covariance – which asserts that the laws of nature can be expressed in the same mathematical form for *all* possible observers in *all* types of motion, so that there is no special class of privileged observers such as those associated with inertial frames in special relativity – was applied by Einstein to a more general type of space-time than that introduced by Minkowski for special relativity. Although Einstein's general theory of relativity involves much more sophisticated mathematics than his special theory, its starting point was a new physical principle concerning the nature of gravitation, first formulated by Einstein in 1907, several years before he made use of the idea of general covariance. The origin of this *principle of equivalence*, as he called it, was his recognition of the far-reaching significance of the fact that, in any small region where gravitational force can be regarded as uniform, all bodies fall with the same acceleration and so are unaccelerated relative to each other. (This is a generalization of Galileo's hypothesis that the Earth's gravitational field imparts the same acceleration to all falling bodies.)

Motion in a uniform gravitational field is therefore equivalent to uniform motion with respect to a frame of reference that has the corresponding acceleration. To illustrate this, Einstein considered the situation inside a *freely falling* lift. Mechanical experience performed inside such a lift will give the same results as if no gravitational force were present and the bodies concerned were weightless – for example, when thrown they will move in straight lines and not in parabolic paths. (Experiments of this kind have actually been performed in recent years, not in a freely falling lift but in space vehicles orbiting the Earth where similar conditions of weightlessness prevail.)

From considerations of this kind Einstein concluded that it is always possible in the case of a *uniform* gravitational field to transform to a space-time frame in which the effects of gravity will not appear. In practice, gravitational fields are not strictly uniform – for

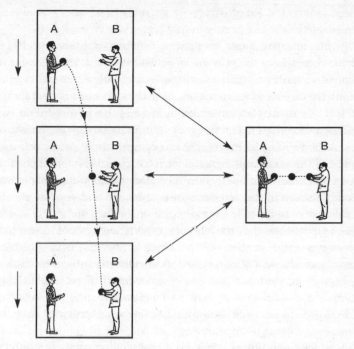

Figure 17: Einstein's principle of equivalence can be expressed as the statement that the effects of gravitation cannot, in themselves, be distinguished from those of acceleration. Two men in free fall, tossing a ball back and forth, will see it travel in a straight line, free from the effects of gravity. But an outside observer, past whom they are accelerating, sees the ball describe a parabolic curve, as it would in a gravitational field.

example, the Earth's field diminishes the farther out in space we go. In the case of non-uniform fields, Einstein's principle has to be modified so as to apply only to very small regions in which the gravitational acceleration varies so little that it is effectively uniform. Thus, Einstein arrived at his final formulation of the principle of equivalence: at any point of space-time it is always possible to choose coordinates so that the effects of gravity will disappear in any region surrounding the point which is so small that the spatial and temporal variation of gravity in that region can be neglected.

In his definitive paper on general relativity, published in 1916, Einstein combined the principle of equivalence with the principle of general covariance applied to a space-time with a general type of geometry known as Riemannian, after the German mathematician G. F. B. Riemann who introduced it in 1854. He also relied on two additional principles. The first was a simplicity postulate: the law of gravitation should be expressible in a comparatively simple way and yet be of the same mathematical form (i.e. covariant) for all possible observers in any kind of motion. The other additional principle was closely related to the principles of equivalence and general covariance and may be called the *principle of the local validity of special relativity*. It states that the laws of special relativity hold locally in a space-time frame of reference (or coordinate system) with vanishing gravitational field. Consequently, special relativity must now imply a negligible gravitational field as the condition of its applicability. Although the velocity of light in empty space may be taken as 300,000 k.p.s. in each *local* frame of reference, provided that the same conventions of measurement are adopted, in general relativity this velocity will not as a rule be a universal constant throughout space-time.

These principles were used by Einstein to obtain a new law of gravitation expressed in the form now known as *Einstein's field equations*. This set of equations relates the gravitational field to the structure of space-time for any system of gravitating bodies. To obtain the equations of motion of a material particle in a particular gravitational field Einstein could not fall back either on Newton's laws of motion or on the modified form of these laws used in special relativity, because these were not compatible with the principle of general covariance. Instead, he made use of a further rule, which was shown by him many years later, working in collaboration with L. Infeld and B. Hoffmann, to be not an independent assumption but a consequence of the field equations. This particular rule was suggested to Einstein by the condition that in special relativity, as in Newtonian theory, a particle moving freely in the absence of external forces moves with uniform velocity in a straight line. In the mathematical reformulation of special relativity by Minkowski, the motion of such a particle is represented, as previously mentioned, by a straight-line path in space-time. This led Einstein to propose that in general relativity the world-line of a particle in a gravitational field should be a geodesic in the space-time associated with this field, a geodesic being the analogue in Riemannian geometry of a straight line in ordinary geometry.

To bring the motion of light within the scope of general relativity, Einstein proceeded in the same spirit as had inspired him to develop special relativity: he extended the principle of equivalence to cover electromagnetic radiation as well as material bodies. He made use of Minkowski's result that in the space-time of special relativity a light-path is one of zero length, and imposed the same rule in the space-time of general relativity. From this condition we can calculate how the gravitational field of a given material system influences the transmission of light and other forms of electromagnetic radiation.

In general relativity Einstein succeeded in absorbing gravitation into the geometry of space-time. For, associated with any system of gravitating bodies, there is a definite space-time (its mathematical description depending on the system of space and time coordinates adopted) such that the gravitational effects of the system are all properties of this space-time. In formulating his field equations Einstein arranged that, in the limiting case of velocities like those of the planets, and of gravitational fields that are of a strength comparable with that of the Sun, the predicted properties of orbits would be very close to those that follow from Newton's theory. This was essential, since it was known that this theory gave very accurate results for the motions of the planets.

Einstein produced three empirical tests of general relativity. First, he was able to account for a small discrepancy between theory and observation in the case of the orbital motion of the planet Mercury that had been known for over thirty years but had completely defeated the efforts of mathematical astronomers working on the basis of Newton's theory. Einstein also predicted, on the basis of general relativity, that the Sun's gravitational field would deflect light rays from a star observed close to the Sun's position in the sky at the time of a total solar eclipse. The precise effect is very difficult to measure accurately, but recently it has been possible to study the same effect in the case of radio waves from a strong extragalactic radio source when occulted by the Sun. The result is close to that predicted by Einstein's theory.

The third of Einstein's tests has the most direct bearing on the properties of time. He found that a gravitational field has a slowing-down effect on natural clocks, analogous to the time-dilatation effect of motion in special relativity. This effect is most readily studied in the spectra of light emitted by a massive body. For the gravitational field of the body will slow down the frequency of this light and so cause its colour to become redder. We call this effect the *gravitational red-shift*. Attempts to detect it in light emitted by the Sun have been made but are difficult because the proportional changes in frequency and wavelength are only of the order of two parts in a million. But, surprisingly, just over ten years ago it became possible to test this effect much more accurately by a laboratory experiment. R. V. Pound and G. A. Rebka of Harvard successfully used a new, highly sensitive technique, called the Mössbauer effect, to measure the change in frequency of light falling from the top to the bottom of a 74-foot tower that had been installed for a different purpose many years before at Yale University. They confirmed with remarkable accuracy the frequency change predicted on the basis of Einstein's theory, which was of the order of only one part in a thousand million million.

These effects are all very small. Generally speaking, in the case of any spherically symmetrical field such as that produced by the Sun the phenomena predicted by Einstein's gravitational theory and Newton's tend to differ significantly only as we get close to the central point. This tendency can perhaps be most simply illustrated in terms of the concept known as 'velocity of escape': depending on how far a particle is from the centre of the gravitational field this is the least velocity the particle can have *at that place* if it is to be possible for it ultimately to move away indefinitely far. In Newtonian theory the nearer one approaches the centre of the field, the larger the velocity of escape becomes, tending to infinity at the centre itself. In relativity theory, however, no material particle can ever attain the local speed of light *c*. In the case of a gravitating spherical body, the distance from its centre at which the velocity of escape is equal to this limiting value is known as the *Schwarzschild radius*: it is given by the formula $2GM/c^2$, where *M* is the mass of the sphere and *G* is the universal constant of gravitation (the same as in Newtonian theory, where the gravitational force between two masses *M* and *m* at a distance *r* apart is GMm/r^2). It can be shown that at this distance from the centre there exists an effective barrier not only for the escape of material particles but also for the escape of light-rays and all other physical signals. No signal from inside this barrier can escape outside, although signals from outside can come in.

It is natural to ask whether such an 'iron curtain', which is peculiar to general relativity with no counterpart in Newtonian theory, could ever actually occur in practice. The solution of Einstein's field equations that involves the Schwarzschild radius is one that applies only in the empty space outside the gravitating body. Inside the body the presence of matter leads to a different solution of the equations, and we find that there is in fact no critical Schwarzschild radius in this case. Consequently, the Schwarzschild radius can only become an actual physical possibility if the body is compressed into so small a volume that it lies entirely inside this radius, which is not the case for any of the bodies in the solar system or for any normal star. For example, the Schwarzschild radius of the Sun is less than 3 kilometres, whereas its actual radius is nearly 700,000 kilometres.

For a body of a mass comparable with that of the Sun or any normal star to be compressed within its Schwarzschild radius it must be of tremendous density, exceeding that of water more than ten thousand million million times. If matter is sufficiently compressed the electrons and protons in it combine to form neutrons, but a body

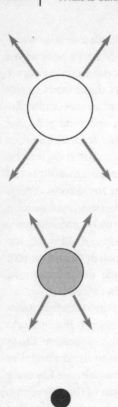

Figure 18:
The velocity of escape – whether of a space-shot, or a particle, or even radiation – from a gravitating spherical body increases as the starting distance from its centre decreases. If the body contracts sufficiently – its mass remaining the same – there comes a stage when the velocity of light would be needed for escape from its surface; after that nothing can escape therefrom, whether material particle or radiation.

composed entirely of neutrons packed as tightly together as possible has a density about a thousand million million times that of water. Until a few years ago there was no evidence that stars of this kind, called *neutron stars*, actually exist. But, since the discovery of pulsars by radioastronomers at Cambridge University in 1968, it seems probable that they do. Pulsars emit radio signals in a regular sequence at intervals of about a second or, in some cases, a tenth of a second. Since no body can emit a sequence of coherent pulses of radiation in a time shorter than would be required for light to travel across it, it was clear that there are pulsars that do not exceed

30,000 kilometres in diameter. Moreover, it can be established theoretically that a period of one-tenth of a second can only be achieved by a body of density at least of the order of a thousand million times that of water. It can also be shown theoretically that a body of this density is likely to contract until it is in the form of a neutron star. Although at present there is no general agreement about the precise mechanism involved, most astronomers believe that pulsars are probably rapidly rotating neutron stars.

So long as white dwarf stars, like the dark companion of Sirius, of density about a million times that of water, were the densest objects known, the question whether any object exists that lies inside its Schwarzschild radius seemed to be unreal; but now that the evidence for neutron stars is so strong, we view this question in a very different light. A body of density greater than ten thousand million million times that of water cannot exist as a neutron star because the gravitational forces acting within it exceed any possible counteracting nuclear forces. Since gravitational force increases with mass, there is a critical mass which a body must not exceed if it is not to collapse under this force. This depends on the density, and for a body of density ten thousand million million times that of water the critical mass is only about 70 per cent that of the Sun's mass. Should a body of greater mass attain this density we know of nothing that can prevent it collapsing under its own gravitational attraction and crushing itself into a singular state of zero radius and infinite density.

The alternative names *black hole* and *collapsar* have been given to what is left behind when a body has suffered gravitational collapse within its Schwarzschild radius. As we have already mentioned, once a body has shrunk within this radius no radiation can emerge nor can matter be ejected. Consequently, to an outside observer, the body becomes completely invisible and no information from it is obtainable. On the other hand, to an observer unfortunate enough to be located on it, complete collapse of the body – and of himself – to zero volume and infinite density would occur in a very short time! To avoid the possibility of this strange fate, it has been suggested that perhaps at very small distances gravitation may cease to be a force of attraction and become a force of repulsion. But if we adhere to general relativity throughout, we are faced with the possibility that there are actually black holes in the universe.

Of all objects that may exist none would seem to offer a poorer prospect of discovery than a solitary black hole, since it would be completely invisible. But if of stellar or greater mass it might be detected by the effect of its strong gravitational field on near-by objects. It has been suggested that a massive black hole may exist at the centre of our Galaxy, and also that in the binary star Epsilon Aurigae the dark component which can only be detected in the infra-red is possibly a dark hole surrounded by a cloud of solid particles from which the infra-red radiation comes. Neither of these suggestions has yet been generally accepted, and it remains to be seen whether the existence of black holes can be conclusively established.

Although gravitational collapse is not a peculiarity of general relativity, since it could also occur for the same high densities on the basis of Newton's theory, there are important differences. In Newton's theory, with its concept of universal time, the collapse to a singularity of infinite density could, in principle, be directly observed from outside. On the other hand, in general relativity the singularity would not only be unobservable from outside, except by its gravitational effects, but the time it would take for the body concerned to shrink to its Schwarzschild radius would be infinitely prolonged *according to an external observer*. This is essentially a time-dilatation effect: as the body shrinks and the gravitational field near it becomes stronger, the time-interval between the reception by the outside observer of signals emitted by the collapsing body at standard proper time-intervals – for example, one second – would become longer and longer, tending to infinity as the body shrinks to its Schwarzschild radius. The most remarkable difference between the two theories in the case of gravitational collapse is that, whereas in the Newtonian theory the singularity would be one affecting matter only, in general relativity gravitation is intimately related to the structure of space-time and consequently the state of infinite density would lead to a singularity in the geometry of space-time. According to the observer associated with the collapsing body this would occur in a finite time. At such a singularity time itself would seem to come to a stop!

A universal cosmic time?

It is not surprising that in recent years a great deal of attention has been devoted by mathematicians to questions concerning space-time singularities. It might be thought that the possibility of these would make it even more difficult to sustain the view advocated at the beginning of this chapter that time must be regarded as a fundamental feature of the universe and that the date is an essential characteristic of any event. For how can this be if a black-hole singularity occurs at a finite time according to an internal observer and in the infinite future according to an external observer?

To resolve this question we must turn to cosmology, that is the science of the structure and evolution of the physical universe as a whole. For, important as the special and the general theory of relativity are for our understanding of time, they do not provide us with a complete account of the concept. These theories are concerned with the nature of physical laws and not with the particular pattern of events that occurs in nature and the actual distribution of matters in the universe. If we regard time as an aspect of the relationship between the universe and the observer, we ought not to conclude that it cannot be a fundamental feature of the universe with objective significance before we have taken due account of the general structure of the universe.

According to current ideas the universe is composed of galaxies, some larger and some smaller, but all, despite certain characteristic differences of structure, roughly comparable with our own Milky Way stellar system. The distribution of these galaxies is somewhat patchy but shows marked signs of general uniformity when considered from a sufficiently large-scale point of view. The discovery of red-shifts in the spectra of galaxies and of their systematic increase as we go to the depths of space has been generally interpreted as evidence that the galaxies are all moving away from each other and that the universe as a whole is expanding.[1] In investigations of theoretical models of the universe, hypothetical observers fixed in the different galaxies are called 'fundamental observers'. These observers are 'privileged' in the sense that they are associated with the bulk distribution of matter in the universe. (The relative velocities of *stars* inside a galaxy are all small compared with the velocity of light, and it is only the relative velocities of minute particles that are more nearly comparable with the velocity of light.) In the models most widely studied the times kept by the fundamental observers fit together to form one common universal time, called *cosmic time*. In other words, according to these observers, there are successive states of the universe as a whole which define a cosmic time. In terms of this, all events have a unique time-order. The anomalies and discrepancies of time-ordering that arise in connection with the special theory of relativity are due not to the nature of events themselves but to the introduction of observers moving through the universe relative to the fundamental observers in their neighbourhood. In general relativity, the strange temporal effects inside black holes concern regions of space-time from which signals cannot be transmitted to any external observer. From the point of view of the fundamental observers there is a common linear time-order for all events that can, in principle, be observed by them.

It can be proved mathematically that the existence of cosmic time in a model universe depends on there being no preferential spatial directions. In other words, if the model is *isotropic*, that is, looks the same in each direction to any fundamental observer, it will possess a cosmic time. In a model of this kind the directions of cosmical recession are like the spokes of a wheel, except that they lie in three-dimensional space and not in a plane, and the general appearance of the universe is the same along each of these

---

1. This question will be considered further in the next chapter.

directions. Moreover, each fundamental observer sees himself at the centre of the same world-picture as any other at the same epoch of cosmic time.

It follows that, to help in deciding whether the actual universe is characterized by a universal cosmic time, we look for corroborative evidence for its isotropy beyond that furnished by the somewhat patchy distribution of the galaxies in the sky. Fortunately, in recent years strong confirmatory evidence has come from an important discovery by radio astronomers. It has been found that the universe is bathed in microwave radiation of a few centimetres' wavelength. Unlike starlight, this radiation appears to be coming to us in more or less equal amounts from each direction of space. The variation in intensity for different directions is actually less than 1 per cent. This degree of isotropy is sufficiently precise to exclude the possibility of any local origin for the radiation, since a source restricted to the solar system, the Galaxy, or even the local cluster of galaxies would not appear isotropic to an observer located, as we are, far from the centre of these systems. This radiation is therefore believed to be a constituent of the universe as a whole and there are strong, although not yet wholly conclusive, reasons for regarding it as a relic of the intensive primeval radiation that may have accompanied an explosive origin of the whole universe. This cosmic microwave radiation is often picturesquely referred to as the 'primeval fireball'. Quite apart from the explanation of its origin, unless we are at a freak centre of isotropy – which seems most unlikely – we must assume that this radiation is isotropic about every fundamental observer in the universe. The evidence for the validity of the concept of cosmic time is therefore now impressive.

Horizons of time

In his *Essay Concerning Human Understanding*, John Locke concluded a chapter on space and time by declaring that 'expansion and duration do mutually embrace and comprehend each other; every part of space being in every part of duration, and every part of duration in every part of expansion'. Until 1917, no one seems to have had cause to regard this statement as anything but a truism. In that year the Dutch astronomer Willem de Sitter constructed a world-model in which time was subject to a curious and previously unsuspected limitation. In the experience of an observer P located at a given point in the model there was a finite horizon at which time appeared to stand still, as at the Mad Hatter's tea-party where it was always six o'clock. This time-horizon was only an apparent phenomenon, since the actual time-flux experienced by any observer Q on this horizon was similar to that experienced by P. This effect occurred because the time required for light, or any other electromagnetic signal, to travel from Q to P was infinite.

In de Sitter's original description of this world-model there was no expansion and no cosmic time. Nowadays, we choose coordinates of space and time so that de Sitter's world is an example of an expanding universe with a cosmic time, its rate of expansion being in fact an exponential function of this time. In terms of cosmic time there is an epoch in the history of B which appears to A to be a time-horizon in the sense that no signal emitted then, or subsequently, by B can ever reach A.[2] Similarly, according to B there is a time-horizon associated with A. A time-horizon of this type is nowadays called an *event-horizon* and will exist for any fundamental observer A in any expanding world-model where the rate of expansion increases with time so fast that eventually signals emitted by B will never arrive at A. As an aid to visualizing this situation, we can picture the universe as the surface of an expanding balloon.[3] The galaxies can then be represented by large dots distributed uniformly over the balloon. One particular dot may be associated with the

---

2. A and B are any two fundamental observers in the model.
3. This analogy must be used with caution. The universe has three dimensions in space, whereas the *surface* of a balloon is two-dimensional. Moreover, in the universe there is nothing corresponding to the regions inside and outside the surface of the balloon.

observer *A*. Light signals can be represented by small dots moving over the balloon with a constant velocity relative to the surface. An event-horizon at *B* will exist for *A* in any world-model where the rate of expansion increases so rapidly that after a certain epoch in *B*'s history no small dot emitted by *B* in the direction of *A* can ever reach *A*, because the rate of expansion has become too fast. In Eddington's graphic phrase, light is then 'like a runner on an expanding track with the winning post receding faster than he can run'.

A different type of time-horizon exists in any world-model which has a decreasing rate of expansion that was at first so fast that no light emitted by *B* could reach *A*, so that only after a certain time has elapsed following the initial state of the model could a signal be emitted by *B* that would eventually reach *A*. In terms of our balloon analogy, initially the balloon expands so fast that its rate of expansion exceeds that of the small dots, so that only after a certain time has elapsed and the expansion has slowed down will a particular *B* become visible to *A*. This type of time-horizon may be called a *creation-horizon*, because it seems to each fundamental observer *A* that matter is continually coming into existence at the confines of the visible universe.

So far we have assumed that *A* remains anchored to a particular galaxy and hence keeps cosmic time, but if he were allowed to move through the universe (with local speed always less than that of light, of course) then the class of events that could, in principle, be observed by him would be increased. If the model possesses an event-horizon for *A* before he moves, then in whatever way he moves he will never be able to observe every event in the universe. His time-horizon will change but can never be abolished.

Most of the expanding models that have been studied as possible forms of the actual universe possess one or other type of horizon or even both. (A notable exception is the uniformly expanding universe considered by E. A. Milne which possesses neither type and so maintains a complete unity in time that is not shared by most models.) Nevertheless, the existence of time-horizons does not invalidate the concept of a fundamental cosmic time associated with the idea of successive states of the universe.

Cosmology and scales of time

The existence of cosmic time does not automatically dispose of the possibility, mentioned at the end of Chapter 4, that different physical laws may define different scales of time that do not keep pace with each other. Given three successive events E, F and G, the temporal intervals between E and F and between F and G, respectively, might be judged to be of equal duration according to one clock and of unequal duration according to another, both clocks being kept by the same observer. Equal intervals of time on one clock will only correspond to equal intervals of time on the other if the mathematical relation between the measures of time on the two clocks is algebraically linear. The two clocks can then be said to keep effectively the same time, any differences between them being adjustable by a suitable choice of time zero and of the unit of time measurement. In the past thirty or more years specific suggestions have been made concerning the possible existence of different natural time-scales that are not linearly related. For example, in 1936 Milne suggested that the time-scale of radioactive decay, in terms of which he believed the universe to expand at a uniform rate, differs from the uniform time of dynamics and gravitation, the latter time being a logarithmic function of the former. Another way of expressing this result was to assert that in terms of the uniform time scale of radioactive decay the universal constant of gravitation G is not a true constant but increases with the cosmic epoch. The geologist Arthur Holmes tried to use this idea to account for the greater activity of the processes beneath the Earth's crust, associated with terrestrial mountain building etc., in the 500 million years that have elapsed since the Cambrian age. He came to the conclusion that the available evidence indicated no significant change in the value of G. Alternative suggestions concerning G and other constants in fundamental physical laws have been made, notably in recent years by R. H. Dicke of Princeton who has modified Einstein's general theory of relativity to allow for a secular change in the value of G.

All suggestions concerning the variability in universal 'constants' such as G agree that the rate of change in its value can be at most a few parts in ten thousand million a year. To detect any such change in a conscionable time we must therefore have some ultra-precise methods of time measurement available. We have already mentioned in Chapter 4 that the accuracy of the caesium clock is about five parts in ten million million. A highly accurate natural time-keeper, which might be checked against the caesium clock, has been discovered in the pulsar. The pulses of radiation emitted by the first of these objects to be detected had a regularity that exceeded one part in a hundred million, the period being precisely 1.33730113 seconds. Although it has since been found that pulsars tend to slow down, it may well be that the periods of some are maintained to a high degree of precision over a sufficiently long interval of time for a partial check to be made of the hypothesis that astronomical time derived from the heavenly bodies keeps step with the time obtained from atomic clocks in terrestrial laboratories. Also it may be possible in due course to compare with great precision the uniform time of celestial mechanics with that of atomic vibrators and of radioactive clocks.

Be that as it may, I believe that until we obtain conclusive evidence to the contrary we should adhere to the hypothesis that there is a unique basic rhythm of the universe. It follows that there is a single universal scale of cosmic time in terms of which, depending on the choice made of time zero and unit of time, every event has, in principle, its own intrinsic date.

7

The origin and arrow of time

THE IDEA THAT ANY EVENT has an intrinsic date leads us to ask how events in the universe can be dated. This is a technical question which has already been considered to some extent in Chapter 4, although the problem of a suitable choice of time zero was not considered. John Locke, in a passage quoted previously, said that 'duration is but as it were, the length of one straight line extended *in infinitum*'. A linear view of time does not, however, commit us necessarily to the conclusion that time is infinite. The dying Hotspur in Shakespeare's *Henry IV*, Part 1, says:

> . . . time that takes survey of all the world
> Must have a stop.

This idea has often been thought of as a gradual 'running down' of the universe – the 'heat death' associated with a direct application of the Second Law of Thermodynamics to the whole cosmos. But, even if we accept this conclusion, and not all scientists do, it is unlikely to provide us with a convenient epoch of reference from which current and past events could be dated. A choice of time zero could, of course, be made by fixing on any particular significant event we like, earlier events being given negative dates and later events positive dates – just as historical events can be given BC and AD dating. If, however, there were a finite origin of time this would be the most natural choice for zero epoch in the cosmic time-scale.

The difficulty that many people have in imagining an origin of time arises because they tend to think of it as being similar to a boundary of space, a concept that we normally reject because of the problem of what we should find on the other side if we got to a boundary; presumably that would be space too and so the concept of a boundary would be self-contradictory. A similar objection in the case of time, however, is not so cogent, since we cannot travel freely in time as we can in space. If time does not exist in its own right but coexists with the universe, an origin (and an end) of time would simply imply a certain temporal restriction on the universe. There is certainly no logical compulsion on us to regard the whole temporal range of phenomena as unlimited. Indeed, if there were any point in so doing, we could regard the universe and all its contents, including simulated human memories, as having come into existence at any particular moment, past or present. Such a moment would be effectively an origin of time.

More realistically, we inquire whether there was some initial singular state of the universe (such as an explosive origin) that can be regarded as a natural origin of time. As a first step towards answering this difficult question we look for evidence of trend in the universe. In a famous lecture on cosmogony in 1871, Helmholtz argued that men of science were not only entitled to, but have a duty to, investigate whether 'on the suggestion of an everlasting uniformity of natural laws, our conclusions from present circumstances as to the past . . . imperatively lead us to an impossible state of things; that is, to the necessity of an infraction of natural laws, of a beginning which could not have been due to processes known to us.' As Helmholtz rightly emphasized, this question is no idle speculation for it concerns the extent to which existing laws are valid.

It was pointed out in Chapter 1 that the decisive factor that led to the prevalence of the linear idea of time was the accumulation of evidence of evolutionary trends in the external world. It is only within the last hundred and fifty or so years that belief in the essentially unchanging character of the universe has been seriously undermined. Until the nineteenth century the concept of evolution made little impact on man's way of thinking about the world. Astronomy, the oldest and most advanced science, did not indicate any evidence of trend in the universe. For, although it had long been realized that time itself could be measured by the motions of the heavenly bodies and that the accuracy of mechanical clocks could be controlled by reference to astronomical observations, the pattern of celestial motions appeared to be essentially the same whether it was read forwards or backwards. The pattern for all motions was a system of wheels and the future was regarded as essentially a repetition of the past. It was therefore not surprising that centuries ago men tended to think of time as cyclic in nature.

The speculations of Kant and Laplace in the latter part of the eighteenth century on the origin of the solar system foreshadowed the nineteenth-century abandonment of the age-old belief that the general state of the world has remained more or less invariable. But it was primarily through the study of fossils in terrestrial rocks that the attention of scientists was turned towards processes of change in the universe that have extended over millions of years. About the year 1800 it seems first to have been realized that the ordering of rocks by strata can be regarded as a chronological ordering. The Darwinian theory of biological evolution, which threw so much light on the nature of the fossil record, was the decisive factor that caused men to become conscious of the time aspect of the universe. Natural selection is a process, often effective only over millions of years, whereby certain genetic combinations tend to be eliminated and others with greater adaptive value are proliferated. The irreversibility of biological evolution is attributed to the comparative improbability of a particular combination of a given set of mutations and a given environment repeating itself, so that the chances of retracing the steps of evolution decrease rapidly with increasing complexity of organisms and environments. According to this view, new mutations lead to new modes of adjustment of organisms to their environments and the subsequent effects of natural selection give rise to that characteristic feature which leads us to think of evolution as a one-way process. As an authority on time direction and evolution, Harold F. Blum, has commented, 'It would be difficult to deny that the over-all effect of evolution is irreversible; the

Ammonites, the Dinosaur and Lepidodendron are gone beyond recall.'

In contrast to the one-way process of biological evolution the surface history of the Earth seems, on first consideration, to be cyclic. The upthrust of land masses from the oceans depends on differential movements in the Earth's crust, influenced at least partially by the flow of heat in the Earth's interior. Since this heat is being radiated away to outer space, a continual source is necessary for the flow to be maintained. This source is now believed to be radioactive deposits in the Earth's crust (p. 69). These deposits are almost sufficient to maintain a steady balance, and consequently the Earth's surface and interior can have remained much as they are now for thousands of millions of years. During the time of the whole fossil record – some 500 million years – there is no evidence of any marked diminution of volcanic and other activity in the Earth's crust, and it has been calculated that heat production by the Earth during this period has not diminished by more than 3 or 4 per cent. Nevertheless, despite the immense period of time during which the cycle of the Earth's surface history has been and is likely to continue to be repeated, a general trend seems inevitable to a steady state in which the continents will all finally be submerged beneath a world-wide ocean.

Turning to the Sun and stars, we find that the processes now thought to be responsible for their continual outpouring of radiation are also essentially irreversible. For, although we now believe that stars like the Sun can continue to shine steadily for thousands of millions of years, the nuclear transformations ultimately responsible for this radiation cannot continue indefinitely. We believe that sooner or later all stars cease to shine in the way the Sun now does and become either white dwarf stars or neutron stars and possibly even black holes. Whatever may be their ultimate fate, the mere fact that stars are radiating bodies that cannot continue in the same state for ever seems to be evidence for unidirectional time.

Hence, both on the terrestrial and on the celestial scale there is abundant evidence of temporal trend *in* the universe, particularly when long intervals of time are considered. But there is also now impressive evidence for associating time *with* the universe. This has come from study of the galaxies. Their spectra are systematically shifted towards the red and this is interpreted as evidence for a general expansion of the universe. This conclusion is based on the combination of two important lines of argument: that the galaxies form the general structure of the universe, and that the red-shifts in their spectra are Doppler effects due to recessional motion.

At first sight, the distribution of galaxies in the sky seems to be extremely irregular. In particular there is a zone of avoidance in the region of the Milky Way. It was shown, however, by Edwin Hubble that this zone of avoidance is due to the existence of diffuse obscuring matter in the outlying regions of our stellar system which prevents us from seeing other galaxies in those directions. When the observed distribution is corrected for this effect, although it is still somewhat patchy, it is much more nearly isotropic. This result, which we have already referred to in Chapter 5, is generally regarded as powerful evidence that the system of galaxies forms the framework of the universe.

As regards the interpretation of the red-shifts in the spectra of galaxies, it is thought unlikely that these are due to the effects of stronger gravitational fields as we go into the depths of space in a static universe, and this explanation has been generally rejected in favour of the Doppler effect.[1] This effect may be briefly described by analogy with the familiar fact that the whistle of a railway engine is shriller when the engine approaches us than when it recedes. This is because the sound waves are more compressed in the former case. Similarly, the light waves from an approaching source are also compressed, and the resulting general shortening of wavelengths means that the spectrum of the source appears to be shifted towards the blue. On the other hand, if the source is receding the light waves from it are elongated and the spectrum is shifted to the red. In each case the fractional displacement (ratio of shift to original wavelength) is the same throughout the spectrum and

---

1. Named after the Austrian scientist who first drew attention to it, in 1842.

provides a means of determining the velocity of the source in the line of sight.

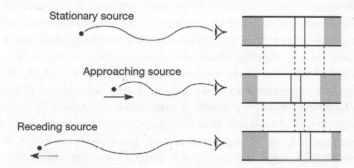

Figure 19: Light waves from an approaching source are compressed, those from a receding one are stretched out. This, the Doppler effect, shifts the wavelength of all the spectral lines of the source towards the short, or blue, end of the spectrum, or towards the red end as the case may be. This shift, if carefully measured, gives the radial velocity with which the source is moving.

Although other interpretations of the displacements in the spectra of galaxies have been advocated from time to time, none has received anything like the degree of support that has been accorded to the Doppler-effect interpretation in terms of recessional motion, and serious objections have been raised against all other theories that specify a precise physical mechanism. Thus, the only viable alternatives would seem to be that either the universe (or system of galaxies) is expanding or else there is an otherwise unknown law of nature operating on the cosmic scale that has no appreciable effect on the scale of ordinary stellar astronomy. Naturally, astronomers tend to favour an explanation in terms of relative motion rather than appeal to an otherwise unknown effect and will continue to do so until there is good reason to abandon this interpretation of the red-shifts.

One of the first to see that the idea of the expansion of the universe could be used to explore the concept of time was the late

E. A. Milne. He argued that, whereas the Newtonian universe *has* a clock, the expanding universe *is* a clock and that time's arrow, to use Eddington's famous expression, is automatically indicated by the recession of the galaxies. In his novel approach to this question, Milne began by drawing an analogy between the expanding universe and a swarm of non-colliding particles moving uniformly in straight lines. If they are contained in a finite volume at some initial instant when they are moving at random, they will eventually form an expanding system. Even if the system is originally contracting, it will eventually expand. On the other hand, an expanding system of this type will never of its own accord become a contracting one. This point can be illustrated even when the system contains only two particles. If initially they are approaching each other, eventually they will be found to be moving apart. But if initially they are moving apart, they will continue to move apart and will never approach each other. This means that, if we took two film strips and found that in one the particles were approaching and in the other receding, we could immediately tell which picture was taken first.

This argument breaks down in a spatially closed universe and therefore presupposes that space is infinite. Similarly, any attempt to define time's arrow in terms of world-expansion presupposes that there is no reversal of the process. It is by no means certain that the universe has never passed through a contracting phase, as would be the case if it alternately expands and contracts in concertina-like fashion. It is therefore preferable to regard world-expansion as a current indication of time's arrow rather than as the basic phenomenon that gives rise to it.

If the universe is expanding then it is likely to have existed in its present state for only a finite time, although the actual measure of this time will depend on whether the recession of the galaxies is approximately uniform or was significantly different in the past. Velocities of recession of galaxies have been discovered up to about two-fifths that of light and there is no reason to regard this as an upper limit. Indeed, in the case of the mysterious quasars red-shifts have been discovered which, if due to recessional motion, indicate speeds of up to nearly nine-tenths of the velocity of light.

In 1929 Hubble found that the velocity v and the distance r of a galaxy are connected by the simple relation $v = Hr$, where H has the same value for all galaxies investigated. The factor H, known as Hubble's constant, has the dimensions of the reciprocal of a time but its accurate evaluation has been impeded by the fundamental difficulties of determining a reliable distance-scale for extragalactic objects. Hubble's pioneer determination of the distances of the nearest galaxies, notably the Andromeda nebula, depended on the identification of constituent Cepheid variable stars, which it was known could be used as distance-indicators. But to obtain a systematic relation between v and r it was necessary to go outside the so-called Local Group of galaxies, and existing telescopes were not powerful enough to reveal stars of this type in more remote galaxies.

In some nebulae, however, notably those in the Virgo cluster, Hubble succeeded in identifying particular objects which he believed, wrongly we now think, to be the brightest constituent stars. By comparing their apparent magnitudes with those of the brightest stars in galaxies whose distances had already been determined, he estimated how far away they, and the galaxies containing them, were. When he correlated the distances and red-shifts of these galaxies, so as to obtain his law, he found that the reciprocal of H was nearly 2,000 million years.

In recent years Hubble's distance-scale of the galaxies has been drastically revised, and this has affected the value assigned to H. The results of the latest investigation by Allan Sandage, announced by him at a conference on the astronomical distance-scale held at the Royal Greenwich Observatory in August 1971, give a value of 18,000 million years for the reciprocal of Hubble's constant, almost exactly ten times that estimated by Hubble himself. If the universe expanded at a uniform rate the law relating distance and velocity would be $r = vt$, where t would be the time that has elapsed since the universe was in an initial state of maximum compression. By comparing the formula with Hubble's empirical law $v = Hr$ (or, to put it in another way, $r = v/H$), we see that in this case t would be exactly equal to the reciprocal of H, which would therefore be a direct measure of the age of the universe. If, however, the universe expanded more rapidly in the past than now, the time of expansion would be correspondingly less; on the other hand, if it expanded more slowly in the past, the time would be increased. If the field equations of general relativity apply to the universe as a whole, uniform expansion is impossible and the simplest world-model is one in which (to use Newtonian terminology) every galaxy is receding with a velocity equal to the velocity of escape from the gravitational field of the whole universe. In this case the rate of expansion continually slows down and the age of the universe is given by two-thirds the value of the reciprocal of H. This comes out to be about 12,000 million years, which is in striking agreement with the age of 10 ($\pm$3) thousand million years estimated for the oldest stars (in globular clusters) on the basis of the theory of stellar evolution.

We must regard this convergence of two completely different lines of research to the same result as either an amazing coincidence or, more plausibly, as strong evidence that some kind of singularity in the past history of the universe occurred between twelve and eighteen thousand million years ago. Although we can take this as a convenient origin of time for the universe as we know it, we cannot rule out the possibility that before the universe was in this singular state – which may have been characterized by an indefinitely high temperature that gave rise to the primeval fireball discussed in Chapter 6 – there was a contracting phase of the universe. The idea of alternate phases of expansion and contraction could mean that the universe passes through an endless sequence of similar cycles. In order to reconcile this idea with the apparently finite life-histories of individual stars and galaxies, it must be assumed that in each cycle stars and galaxies are created anew out of whatever remains in existence from the previous cycle. We know of no way in which this could happen, and our present knowledge compels us, unlike thinkers in previous civilizations whose speculations on cycles of time and the universe were more in the nature of poetry than science, to confine our attention to what is happening in the current cycle. Even here we are up against a serious difficulty. For if the universe is cyclic and not continually expanding, there are difficulties in understanding the physics of the contracting phase that must eventually set in, when all extragalactic red-shifts become blue-shifts. Nevertheless, we cannot yet automatically exclude such an oscillating type of universe from the realms of physical possibility.

The thermodynamic arrow

We have seen how the idea of a natural origin of time can be associated with a singularity in the past history of the universe. Such a singularity is implied by the most generally accepted interpretation of the red-shifts in the spectra of the galaxies. The close association of time with the universe is therefore revealed by the history of the *material contents of the universe*. On the other hand, at the end of Chapter 4, an alternative suggestion was made that the ultimate scale of time is closely associated with our concept of *universal laws of nature*. In that chapter we were only concerned with the measurement of time and not with the problem of dating and the associated questions of the choice of time zero and of the direction of time's arrow. When we seek corroborative evidence from our knowledge of physical laws to support the answers to these questions arrived at by our investigation of the distribution of matter on the astronomical scale, we are confronted by a strange situation. So far from obtaining any such evidence, we find that both the laws of motion and the laws governing the known forces and interactions in physics (with one minor exception) are compatible with the reversal of time's arrow! This is true for the laws of gravitation (both Newton's and Einstein's), of electromagnetism and of those governing the so-called 'strong' interactions between protons and neutrons in atomic nuclei. Only in the case of the 'weak' interactions, involving neutrinos, has any doubt been cast on time-reversal invariance. Apart from this one possible loophole, it does not seem that the laws governing the fundamental forces and interactions in nature provide any indication of time's arrow.

Whatever the true situation as regards weak interactions, it has long been thought that evidence for time's arrow should be sought in the statistical aspects of nature. The most celebrated physical law relating to these aspects of the world is the Second Law of Thermodynamics. It is of a quite different character from the laws governing natural forces. It is concerned with the temporal behaviour of large numbers of particles and has no meaning for individual particles or systems containing few particles. It is essentially statistical and asserts the tendency of orderly arrangements of molecules to break down into disorderly arrangements. A familiar example from everyday life is the effect of stirring when cream is poured into a cup of coffee. In a short while we obtain a liquid of uniform colour, and however long we stir we never find the contents of the cup reverting to their original orderly state in which coffee and cream were clearly separated.

The Second Law of Thermodynamics was interpreted by the Austrian physicist Ludwig Boltzmann as the statement that any closed system, that is to say one that neither gains energy from nor loses energy to its environment, automatically tends to a state in which the distribution and motion of its constituent parts is completely random – since for a large number of parts this is much the most probable arrangement. This is what happens when the contents of a cup of coffee are well mixed. It is most unlikely that continued mixing will produce an arrangement in which coffee and cream become clearly separated again. Boltzmann suggested that this statistical interpretation of the Second Law automatically accounted for the directional character of time itself.

Despite its apparent cogency, this conclusion was found to be logically unsound. For, owing to the symmetry of the laws of motion with respect to both directions of time, it follows that for any arbitrarily chosen state that shows some degree of ordering there is not only a large probability that it will lead to a less orderly state, but equally there is a large probability that it itself arose from a disorderly state. It would therefore seem to be highly likely that at the time of the chosen orderly state the system is undergoing a fluctuation from disorderliness. Consequently, the statistical theory of Boltzmann does not provide an unambiguous direction for time's arrow, which must therefore be accounted for – if it can be explained at all – in some other way.

However, the curious situation that the general laws of physics (both on the sub-atomic and the macroscopic scales) cooperate to conceal the one-way trend of time from us must not lead us to conclude that time's arrow is a purely subjective phenomenon, as has been suggested by some philosophers. In order to apply the laws to a given physical system we impose certain contingent 'initial conditions', such as the positions and velocities of constituent parts of the system at some chosen instant. These conditions are not given by the laws themselves, which are of a general, and not a particular, character. If there is some deep connection between time and the universe, this may be because time's arrow is associated in some way with the 'initial conditions' that determined the particular universe that actually is, as distinct from any other universe which might have existed in accordance with the same physical principles. In other words, it may be that the ultimate explanation of time's arrow will be found in cosmology, that is in the particular pattern of events in the universe, rather than in general laws.

The three arrows of time

Recently David Layzer of Harvard has made a fresh attempt to relate the three macroscopic arrows of time: the thermodynamic arrow, defined by entropy[2] processes in closed systems, the historical arrow, defined by information-generating processes in certain open systems,

---

2. Entropy is a measure of the disorganization of a physical system. The Second Law of Thermodynamics implies that in a closed physical system, not subject to interference from outside, entropy automatically tends to increase.

and the cosmological arrow, defined by the recession of the galaxies. Having pointed out the subtle difficulties associated with the thermodynamic arrow, he turns to the historical arrow provided by the evolutionary records which all point in the direction of increasing information. These records are produced not only by biological systems. A record of the Moon's past is written in its pitted surface; the internal structure of a star, like that of a tree, records the process of ageing; and the complicated forms we observe in spiral galaxies reflect the evolutionary processes that shape them. We may define the historical arrow through the statement that 'the present state of the universe (or of any sufficiently large subsystem of it) contains a partial record of the past but none of the future.'

Layzer sketches a theory that seeks to relate the thermodynamic and historical arrows with the cosmological one by deriving all three from a common postulate: that the spatial structure of the universe is *statistically* homogeneous and isotropic – in other words, no statistical property of the universe serves to define a specific direction or position in space. From this he deduces that a *complete* description of the universe can be expressed in statistical terms. For example, if a universe satisfying Layzer's postulate were in a state of thermodynamic equilibrium it would be completely characterized by its temperature and density, and all other observable quantities could be calculated knowing only these. Layzer argues that thermodynamic equilibrium of the whole universe is only likely to be satisfied when the universe is close to a singular state of infinite density, which he defines as the initial state. The cosmic expansion then generates both entropy and information. He concludes that the world is unfolding in time and the future never wholly predictable, since the specific information content of the universe increases steadily from the initial singular state. Consequently, the present state of the universe cannot contain enough information to define any future state. 'The future grows from the past as a plant grows from a seed, yet it contains more than the past.'

Despite the ingenuity of attempts such as Layzer's to account for time's arrow and the valuable insight which arguments such as his provide into the relationships that prevail between the different one-way processes in the universe, any theory that seeks to derive the *entire* concept of time from some more primitive considerations – for example, assumptions of a causal, probabilistic or statistical nature – is foredoomed to failure. For any theory which endeavours to account for time *completely* ought to explain why it is that everything does not happen at once. Unless the existence of *successive* (non-simultaneous) states of phenomena is tacitly assumed it is impossible to deduce them. What the many attempts to analyse the nature of time have shown is that ultimately time must be regarded cosmologically. In the final count, time is a fundamental property of the relationship between the universe and the observer which cannot be reduced to anything else.

8

The significance of time

THE CONCEPT OF TIME has never ceased to intrigue and puzzle those who think about it. We feel that, whatever happens, time must go on unceasingly and yet, when we come to analyse it, we find good reasons for rejecting the idea that time exists in its own right. We regard time as the order in which events happen. Consequently, if there were no succession of events, there could be no time. 'What did God do before He made Heaven and Earth?' asked St Augustine, He rejected the facetious answer that God was preparing hell for those who pry into mysteries! St Augustine's answer was that before God made heaven and earth he did not make anything. Time was made along with the heaven and the earth.

The intimate association of time and the universe can be traced back to Plato, the philosopher who exercised such a great influence on St Augustine. In Plato's cosmology, as set out in his dialogue the *Timaeus*, the universe was fashioned by a divine artificer who imposed order on primeval chaos by reducing it to the rule of what nowadays we call natural law. In Plato's view, the pattern of law was provided by ideal geometrical shapes in a state of absolute rest, and therefore essentially *timeless*. Whereas space was regarded by Plato as a pre-existing framework into which the universe is fitted, time was itself produced by the universe. For the universe, unlike the eternal ideal model on which it was based, is subject to change, and time is that aspect of change which bridges the gap between the two (the material universe and its ideal model), being, in his famous phrase, the 'moving image of eternity'.

Plato's antipathy to all research that involved the temporal material world led him to criticize people like the Pythagoreans who investigated problems of musical harmony and acoustics empirically. In an amusing passage in the *Republic* he pokes fun at them for wasting their time in measuring audible sounds and concords.

> They lay their ears to the instrument as if they were trying to overhear the conversation from next door. One says he can still detect a note in between, giving the smallest possible interval, which ought to be taken as the unit of measurement, while another insists that there is now no difference between the notes. Both prefer their ears to their intelligence.

Plato was convinced that 'heard melodies are sweet, but those unheard are sweeter' – an attitude that was deeply influenced by the work of an earlier philosopher, Parmenides, the founding father of strict deductive argument and logical disputation. Parmenides submitted the ideas of becoming and perishing to acute criticism and concluded that time does not pertain to anything that is truly 'real', but only to the logically unsatisfactory world of appearance revealed to us by the senses.

Parmenides' belief that temporal flux is not an intrinsic feature of the *ultimate* nature of things has been tremendously influential. It is not only idealist philosophers who have claimed that the temporal mode of our perception has no ultimate significance. Even so empirically-minded a thinker as Bertrand Russell, although he rejected the arguments by which these philosophers have sought to justify this conclusion, made the following admission in his well-known essay on 'Mysticism and Logic': 'There is some sense – easier to feel than to state – in which time is an unimportant and superficial characteristic of reality. Past and future must be acknowledged to be as real as the present, and a certain emancipation from slavery to time is essential to philosophic thought.' Alas, even philosophers are men like the rest of us. An amusing story is told of the Russian philosopher Nicolas Berdyaev, who, after he had pleaded passionately for the insignificance and unreality of time, suddenly stopped and looked at his watch with genuine anxiety, fearing that he was a few minutes late for taking his medicine!

It is notorious that most philosophers have regarded time as a thoroughly unsatisfactory concept. The French psychologist Pierre Janet remarked some forty years ago, in his book on time and memory, that, whenever stress is laid on logic and reason, time tends to be unpopular. Philosophers usually have a particular horror of the concept and have done all they can to suppress it. Nevertheless, it is only fair to point out that many mathematicians and physicists, too, have been sceptical about the ultimate significance of time and have been far more favourably inclined to spatial concepts. To some extent this may be because space seems to be presented to us all of a piece, whereas time comes to us bit by bit. The past must be recalled by dubious aid of memory, the future is unknown, and only the present is directly experienced. Even Einstein, who made the greatest contribution since the seventeenth century to the understanding of time when he formulated his special theory of relativity, later became decidedly wary of the concept, as we have seen, and came to the conclusion that physical reality should be regarded as a four-dimensional existence rather than as the *evolution* of a three-dimensional existence. In other words, the passage of time is to be regarded as merely a feature of our consciousness that has no object-ive physical significance. This sophisticated hypothesis makes the concept of time completely subordinate to that of space.

But time has certain important characteristics that clearly distinguish it from space. Apart from its one-dimensional nature, the two principal features peculiar to our conception of time are its arrow and its passage. Whereas time's arrow depicts the irreversible before-and-after succession of events, time's passage refers to the distinction that we make between past, present and future. These two closely associated properties must not be confused.

The before-and-after series is a permanent series in the sense that, if the statement '*B* occurs after *A*' is true, it is always true. For example, the statement that the Battle of Waterloo occurred after the Battle of Hastings is a permanent truth. The before-and-after series is the way in which we normally *contemplate* a chain of events in time. It is a method of ordering analogous to numerical ordering and is compatible with the 'block universe' idea. On the other hand, the series of past, present and future characterizes the way in which we actually *experience* events. Unlike the before-and-after series, it is a changing series and gives meaning to the concept of occurrence. The fact that it is a changing series – that what happens now was once future and will be past – leads us to make statements that are not permanent truths. For philosophers this changing series has frequently been such a source of perplexity that many of them have concluded that it must be an illusion – a view that was held by the Cambridge philosopher M. J. E. McTaggart. The foundation of his argument was his contention that an event can never cease to be an event. 'Take any event,' he wrote, 'the death of Queen Anne, for example – and consider what changes can take place in its characteristics. That it is a death, that it is the death of Anne Stuart, that it has such causes, that it has such effects – every characteristic of this sort never changes.' McTaggart argued that from the dawn of time the event in question was the death of a Queen. He went on:

> At the last moment of time – if time has a last moment – it will still be the death of a Queen. And in every respect but one, it is equally devoid of change. But in one respect it does change. It was once an event in the far future. It became every moment an event in the nearer future. At last it was present. Then it became past and will always remain past, though every moment it becomes further and further past.

McTaggart argued that, although past, present and future are incompatible, they must apply to every event. One might make the obvious retort that events do not have these characteristics simultaneously but successively, in which case McTaggart could easily counter with the argument that our statement that an event is present, will be past, and has been future, means that the event is present at a moment of present time, past at a moment of future time, and future at some moment of past time. But each of these moments is itself an event in time and so is both past, present and future: in other words, the difficulty breaks out all over again and we are launched on a vicious infinite regress. McTaggart concluded that time is an illusion. This conclusion is, in my opinion, a non-sequitur. McTaggart's error was to treat the happening of events as if it were a form of qualitative change. But time is not itself a process in time.

Although in recent years few people have been influenced by McTaggart, a number of philosophers and others have argued that the transitional aspect of time is purely subjective. They do not regard it as a characteristic of physical time itself but only of our perception of time. In particular, they claim that our concept of 'present', which we signify by the word 'now', is merely the temporal mode of our personal experience, so that if there were no such experiences there would be no 'now'. This point of view is adequate for a great deal of physics and other sciences, so long as dates are irrelevant and the particular time at which an experiment is performed does not matter. In such cases, when classifying events temporally it is sufficient to concentrate on the relations of 'earlier than', 'later than', or 'simultaneous with'. On the other hand, for the meteorologist engaged in forecasting the weather, the precise distinctions between past, present and future are vital. Similarly, for the palaeontologist studying the fossil record in terrestrial rocks, not only are dates relevant but the distinction between past and present dominates his thoughts, since the overall effect of evolution appears to be irreversible.

Nevertheless, some philosophers have argued that one cannot define the present except by reference to itself. The present, they would argue, is simply our 'now', and as this is a circular definition there is no reason to suppose that what it defines has objective significance. They therefore conclude that we should restrict the concept 'now' to our mode of perception. Instead of accepting this view, can we establish the objectivity of past, present and future?

First, let us consider what we mean by simultaneity and the present. In the terminology introduced into relativity theory by Minkowski, two events on the respective world-lines of two distinct individuals *A* and *B*, whether living or inanimate, are simultaneous if they are located at a point of intersection *O* of their world-lines. To establish the objectivity of a phenomenon, we usually try to show that it is not just a peculiarity of a particular person's experience. For example, one fine night in November 1572, the famous Danish astronomer Tycho Brahe, who knew the starry skies like the back of his hand, saw to his surprise a bright star (it was in fact what we now call a 'supernova') where no other star had ever been seen before. His doubts concerning its objective existence were resolved when he found that other people (his servants and some peasants driving by) saw it, too. Similarly, if the concept of the present is objective, any *A* and *B* when at the same *O* must have the same 'now'.

What would it mean if *A* and *B* had different 'nows' when they are simultaneously at *O*? Since, for this purpose we cannot usefully compare any purely internal feelings of presentness, because they are subjective and we are seeking to establish objectivity, we must concentrate on external physical events and the relation of the individual to the environment. An individual's 'present' can be defined as all that which interacts with him at a given instant. It is a relation between an individual and the rest of the universe, being all that which is happening to him at a given instant – all that which is in fact present for him. This definition does not necessarily imply self-awareness and can be applied to any individual, inanimate as well as animate, so long as it is capable of interacting with its environment.

Having defined the present in this way, can we show that it is an objective concept? Clearly, the only point in claiming the opposite would be if two individuals (animate or inanimate) could simultaneously have different 'nows'. This would happen if when *A* and *B* are together at *O* they were to have incompatible interactions with their environments. Suppose *A* is a mirror capable of reflecting light that falls on it and *B* is a human being. If *A* were to present a view of trees in winter without foliage when B sees the same trees in full leaf, we could interpret the discrepancy as evidence that *B*'s 'now' is out of phase with *A*'s.

In practice, we do not normally[1] encounter this kind of discrepancy, and the physical world would be much more complicated if we did. We have therefore no reason for rejecting the commonsense assumption that *A* and *B* have a common 'now', and from this it follows that the distinctions we make between past, present and future are not merely subjective.

It has already been shown that acceptance of the theory of relativity does not compel us to regard the order of events in time as wholly dependent on the observer. For, as we have seen, the theory actually allows an objective time-order for a wide class of events: namely those which can interact with or influence each other. Consequently, in defining the concept of the present for any observer in terms of his interaction with his environment we are not in conflict with relativity. Moreover, if the universe admits a common cosmic time for observers fixed in the galaxies, then in terms of this cosmic time all events have a unique time-order. From the point of view of the fundamental observers there is a common linear world-time time-order and a clear-cut distinction between past, present and future. We come back to Plato's idea that time and the universe are intimately associated.

Atoms of time

The concept of time that has prevailed in the last few centuries is based on the idea of linear advancement, but it also assumes that time is homogeneous and continuous. The plausibility of these assumptions was not only greatly strengthened by the development of precise methods and machines for the measurement of time, but also more subtly by the general decline of belief in traditional temporal associations of a magical rather than a scientific nature. It is true that the notions of lucky and unlucky days and of climacteric

---

1. Hallucinations, optical illusions, etc. being ruled out as fake evidence, and allegedly paranormal phenomena rejected.

years – those periodic dates in a man's life which were potential turning-points in his health and fortune and were based on the doctrine that a man's body changed its character every seven years – were all rejected by the medieval Church. But the ecclesiastical calendar also tended to encourage belief in the uneven nature of time.

In the seventeenth century many of the traditional practices enshrined in this calendar, such as the observance of Lent and the celebration of saints' days, were attacked by the Puritans, who advocated instead strict adherence to a regular routine of six days of work followed by a day of rest. By the end of the seventeenth century this routine had become generally accepted in this country. According to Keith Thomas, who has made a thorough investigation of popular beliefs in sixteenth- and seventeenth-century England in his book *Religion and the Decline of Magic*, this change in working habits was 'an important step towards the social acceptance of the modern notion of time as even in quality, as opposed to the primitive sense of time's unevenness and irregularity'. Nevertheless, a relic of this more primitive conception of time survived in the strict observance of Sunday as a day of rest that was still rigorously enforced in many households within living memory.

Belief in the unevenness of time was more natural in the past when society was essentially agrarian and dependent on the seasons for its pattern of living. The medieval Christian almanac, with its emphasis on the *year*, was based on the needs of this type of society, whereas the Puritan insistence on a rhythm of living based on the *week* was more natural to those who worked in towns instead of being tied to the soil. During the latter part of the seventeenth century developments in economic life began to prevail over the traditional seasonal routine and this made for general acceptance of the scientific idea of homogeneous and continuous time.

Nowadays most of us tend to accept automatically the idea that time is continuous because we believe in the continuity of our existence. Until the present century it was also possible to believe in the continuity of matter and energy, but with the establishment of the atomic theory of matter and of the quantum theory we have been compelled to abandon these beliefs. In recent years the continuity of time has occasionally been called in question, although in this case it is too soon to say what the ultimate decision will be. Instead of time being infinitely divisible it has been suggested that, like matter and energy, it may be atomic or granular in structure. This speculation is linked with a similar idea concerning the nature of space. It has been claimed that a minimum spatial displacement may be about a million millionth part of a millimetre (corresponding to the effective diameter of a proton or electron). If this is so, then a corresponding minimum time – the *chronon* – might be the time required for light (the fastest moving thing) to cross such a distance. This would be about the million millionth part of the million millionth part of a second (10^{-24} sec). Of course, if the chronon exists it would be a minimum value of proper time and owing to time dilatation would appear relatively shorter to a moving observer.

If time does consist of a sequence of 'atoms' as short as this, it would, for all practical purposes, be virtually continuous. Nevertheless, from the theoretical point of view, however small its magnitude may be, the possible existence of the chronon is a revolutionary idea that calls into question a fundamental feature both of the scientific idea of time that has prevailed in recent centuries and of the popular conception of time that most people accept intuitively.[2]

Precognition and the nature of time

Another of the traditional properties of time that has also occasionally been called in question in recent years is its unidimensionality. Some investigators of extrasensory perception have argued that

---

2. Many people have difficulty in imagining time to be 'atomic' in structure because they believe that this would imply the existence of temporal gaps which must themselves be a part of time in contradiction of the hypothesis. On the contrary, however, the 'atomicity' of time refers only to the *indivisibility* of the chronon. In principle, chronons could be imagined as being like a row of pebbles which touch each other, so that there would be no gaps between them.

linear time is inadequate to account for all events in our world. In particular, the idea that time may have more than one dimension has been invoked by J. W. Dunne, in his well-known book *An Experiment with Time*, to justify his claim that occasionally in dreams future events in our waking life are experienced as pre-presentations. An event *P* might precede an event *Q* in the familiar time-dimension and yet *Q* might precede *P* in another time-dimension. Consequently, if *P* were the precognitive impression of event *Q* – for example, a dream concerning the occurrence of *Q* before *Q* actually occurs – it would be intelligible to say that *Q* determines *P* if it occurred before *P* in the second dimension of time.

Any hypothesis of this kind involving a second dimension of time is difficult to accept because it means that we should have to cope with the puzzling notion of a double 'now', for what is 'now' in one respect, or dimension, could be 'past' or 'not yet' in the other. Worse still, it would lead to the following curious situation.

Suppose I precognize an event which is to occur next Sunday. In one respect – that is to say, in one dimension of time – this event has not yet come into being: it is still future and does not yet exist. But in another respect, or second dimension of time, it is past and so has already come into being. It is, so to speak, half-real, since it has partially come into existence but not entirely. Not until next Sunday will it receive its second instalment of being, and then be completely real. But will it? For these two parts of its being are in fact out of step, because what begins to be in one dimension of time will already be long past in the other.

The possibility of precognition has been rejected by the Cambridge philosopher C. D. Broad who argued that the phrase 'future event' does not describe an event of some special kind, as the phrase 'sudden event' or 'historic event' does. Instead, a future event is nothing but an unrealized possibility until it comes to pass and therefore can itself influence nothing, although the present *knowledge* that there will be such an event can influence our actions when it is called to mind. An event which seems to 'fulfil' an earlier experience and make it appear precognitive cannot possibly help to determine the actual occurrence of that experience. Consequently although there may be cases of apparent precognition, they cannot be genuine pre-perceptions, and the hypothesis of two-dimensional time is certainly not required to explain them.

The transitional nature of time

Genuine precognition, in the sense of our being able, in certain circumstances, to perceive future events before they actually happen, might perhaps be possible if we inhabited a block universe in which, as I said before, physical events do not suddenly occur but are there waiting for us to experience them. This idea has already been rejected on the grounds that past, present and future are in fact objective characteristics of physical events. But the block-universe hypothesis has strange implications for mental events, such as our conscious perceptions and our decisions to perform physical actions. In a block universe, as we have seen, past, present and future do not apply to physical events, and so they neither come into existence nor cease to exist – they just are. But whatever kind of universe we inhabit, mental events certainly come to be and cease to be in our personal experience. Therefore, if we inhabited a block universe, mental events would have a completely different kind of existence from physical events. This would have the most peculiar consequences for cause and effect. In purely physical causation, an effect would not actually be produced by its cause, it would merely be further on in time. But mental causation of a physical event – such as deciding to drop a stone into a pond – would mean that a cause (in this case, the decision to drop the stone) suddenly comes into being, but the effect (the splash when the stone strikes the water) would not: it would just be. Such a strange difference between cause and effect would be completely incomprehensible.

If physical events are there eternally, how could we get the illusion that they are not? Surely, we have the faculty for temporal awareness of successive phases of sensory experience because our minds are adapted to the world we live in and this is a constantly changing world. The objections that are brought against the transitional nature of physical time are a rearguard action in favour of the age-old belief in the essentially unchanging character of the universe and the ultimate insignificance of its temporal aspect. So far from the transience of time being an inessential, because purely subjective, characteristic, the ultimate significance of time is to be found in its transitional nature. For, just as the reason for the existence of evil in a moral universe must be that without evil there could be no good – since there would then be nothing to contrast good with and thereby give meaning to the concept – so without the fact of transience there could be no significance in permanence.

To conclude: although our perception of time has many subjective and even sociological features, it is based on an objective factor that provides an external control for the timing of our physiological processes. This objective factor is what we call physical time. It is an ultimate feature of the universe and its relationship with observers, particularly fundamental observers, which cannot be reduced to anything else. But this does not mean that it exists in its own right: it is an aspect of phenomena. The essence of time is its transitional nature. That this has given rise to so much argument down the centuries is not surprising, for, in the words of Whitehead, 'it is impossible to meditate on time and the mystery of the creative passage of nature without an overwhelming emotion at the limitations of human intelligence'.

APPENDIX

Temporal order in special relativity

To establish the results discussed on page 90 appeal must be made to the Lorentz formulae of special relativity. These formulae are given in any textbook dealing with the subject.

The first observer A and the second observer B are together at event E, to which each therefore assigns zero distance from himself. Also each observer can arrange to set his clock so that it reads zero epoch at this event. To the event F which occurs at distance r from A, suppose the epoch assigned by A is t. Since, according to A, the event F occurs later than E, the value of t is greater than zero. If the velocity of B relative to A is V, then according to the Lorentz formulae the time t' assigned by B to the event F is not the same as t, as it would be in classical Newtonian physics, but is given by

$$t' = \frac{t - Vr/c^2}{\sqrt{(1 - V^2/c^2)}}$$

where c is the velocity of light.

Since V must be less than c, it follows that if the distance r, according to A, of the event F is less than ct, then t' is greater than zero. Consequently, B agrees with A that F occurs later than E. Similarly, if $r = ct$, we find that t' is still greater than zero for all permissible values of V, and so B again regards F as occurring after E.

If, however, r is greater than ct, we can find a permissible value of V for which $t' = 0$, namely $V = c^2 t/r$, which in this case is less than c. Consequently, according to the particular observer B moving with this velocity V relative to A, the event F is simultaneous with E.

Moreover, if V is chosen to be greater than $c^2 t/r$ (but less than c of course), it follows that t' will be negative. Hence, in this case the observer B regards the event F as occurring *before* the event E.

These results, which have been established algebraically, can be illustrated geometrically by the following diagram. The axes represent times t and distances r according to A. Any event to which B assigns the same time $t' = 0$, as he does to E, is represented by a point on the line marked $t' = 0$, for which $t = Vr/c^2$, according to A. This line always lies below the line $t = r/c$, because V must be less than c. If r is less than ct, then F will lie above the line $t = r/c$, say at F_1, and so above the line $t' = 0$. Consequently, the value of t' assigned by B to F is greater than zero, the time assigned to E. If, however, r is greater than ct, then F lies below the line $t = r/c$, say at F_2. We can then find an admissible value of V for which F_2 lies on the line $t' = 0$, and so for the corresponding observer B the event F_2 is simultaneous with E. Finally, if F lies below this line, say at F_3, then t' is less than zero and B regards the event F as occurring before E.

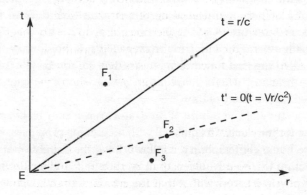

THE LITERATURE OF TIME*

J. T. Fraser and M. P. Soulsby

In Shakespeare's *Henry IV*, Part 1, Sir Henry Percy appears as a restless young man. He is nicknamed Hotspur because of his diligent patrolling of the border between Scotland and England. He is fiery tempered and quick on the draw. But when he is about to die, he becomes reflective. 'Time that takes survey of all the world', he says, 'must have a stop.'[1]

The debate about an end of time has been going on for a long while, without having reached a universally acceptable conclusion. In contrast, there is universal agreement that time does survey — time does enter into — all negotiations of people with themselves, with each other, and with their environments. It would then follow that everything that has ever been experienced, noted, said, or created by humans, whether fleeting or lasting, should be considered material for the study of time.

Yet, a study of the nature of time, and even a thin bibliographic sampler for the same, if formulated without guiding principles, would become but a contemporary version of Dickens' delightful Mudfog Association for the Advancement of Everything.[2] And as incomplete as the physical theories-of-everything which, as J. D. Barrow ably remarked, are 'far from sufficient to unravel the subtleties of a

---

* Dedicated to the members of the International Society for the Study of Time.
1. *1 Henry IV*, V. iv. 82.
2. The Mudfog Association is the subject of Charles Dickens' *Sketches by Boz*, New York: Bigelow, n.d. Vol. 2, 'The Mudfog Papers', pp. 353–436.

Universe like ours.'[3] There is a need for selection rules in the compilation of bibliographic entries on any subject. Such a need, one for determining what is and what is not relevant to a field of knowledge, is hardly unique to the study of time. Not every line drawn by ape or man is necessarily an example of art, to be so honored. Although all such lines are subject to the rules of geometry, not all of them need to be studied for the formulation of a science of geometry. Thus far, in the integrated study of time, judgments about the relevance of material depended on the practitioners of those disciplines to which appeals were made. Let the botanist decide what is important for an understanding of the cyclic and aging behavior of plants. How to appeal to specialist judgments through extensive critical notes and references is illustrated by G. J. Whitrow's classic, *The Natural Philosophy of Time*. How it may be done in an informal manner is illustrated by the present volume.

What attempts have been made, thus far, for a systematic survey of material important for an integrated study of time?

A 1981 computer survey considered 15 databases, containing an estimated fifteen million books and articles published between 1965 and 1980. Works of possible interest published between 1900 and 1964, not then available in computerized databases, were identified from library cards. The databases searched, the subjects sought, the strategy of the search, and the guidelines of selection are described in J. T. Fraser's 'Report on the Literature of Time, 1900–1980'.[4] The total number of entries judged to be potentially relevant to a systematic study of time was estimated as approximately 65 000.

Samuel L. Macey's *Time: A Bibliographic Guide*[5] contains about 6000 entries on time-related books and articles. They are divided into 25 academic disciplines and about a hundred subdisciplines.

The 10 volumes of *The Study of Time* series (1972–2001) contain

---

3. Barrow, J. D. *Theories of Everything: The Quest for Ultimate Explanation*. Oxford: Clarendon Press, 1991, p. 210.
4. Fraser, J. T., Lawrence, N., and Park, D. (ed.). *The Study of Time IV*. New York: Springer-Verlag, 1981, pp. 234–70.
5. New York: Garland, 1991.

over 230 papers, each carrying extensive notes and references. A listing of these papers – by authors, titles, and locations – may be found in *KronoScope: Journal for the Study of Time*, Vol. 2, Issue 2 (2002), pp. 263–73. The 'Time's Books' column of *Time's News*, the aperiodic newsletter of the International Society for the Study of Time (<http://www.studyoftime.org>) carried, between 1984 and 2002, over 300 book reviews. That column is now a part of *KronoScope*.

Before a defensible scheme of criteria for the selection of material relevant to an integrated study of time, together with a critical, classificatory scheme for works, may be formulated, it will be necessary to construct a 'coherent, intellectual structure' and identify 'a vocabulary of concepts' called for in the introduction to this book, 'On ye sholders of Giants'.

This sampling begins with a selection from the time-related works of G. J. Whitrow. It continues with titles suggested, for this volume, by members of the International Society for the Study of Time. To them, the compilers wish to express their appreciation. The entries are arranged under the eight chapter headings of this book. Each entry was placed in what was judged to be the best, even if not perfect fit. Some edited volumes, however, do not fit easily or appropriately in the classification guided by the chapter headings of this book. They may provide interdisciplinary approaches to the study of time or a specialized focus. A section for such edited books has been included to allow for the variety and scope of material. The sampling concludes with titles of those works of J. T. Fraser that specifically address problems of the integrated study of time.

The final judgment for the relevance and value of any and all of the entries listed below remains with the reader. As Lady Macbeth's physician remarked, 'Therein the patient /Must minister to himself.' Or herself.

A SELECTION FROM THE TIME-RELATED PUBLICATIONS OF
G. J. WHITROW

Books

The Natural Philosophy of Time. Oxford: Clarendon Press, 1961; 2nd
    edn, 1980.
*Time in History: The Evolution of our General Awareness of Time and
    Temporal Perspective*. Oxford University Press, 1988.
What is Time? London: Thames & Hudson, 1972.

Articles

'Albert Einstein: a biographical portrait'. BBC, transmitted on 30 July
    1966.
'The concept of time from Pythagoras to Aristotle'. *Proceedings of the
    10th International Congress for the History of Science*. Paris, 1964,
    pp. 499–503.
'Entropy'. In *The Encyclopedia of Philosophy*, Vol. 2 (ed. Paul Edwards).
    New York: Macmillan, 1967, pp. 526–8.
'General Relativity and Lorentz invariant theories of gravitation'.
    Nature, **188**, 3 December 1960, 790–4.
'Man and time: some historical reflections'. In *The Study of Time VI*
    (ed. J. T. Fraser). Madison, CT: International Universities Press,
    1989, pp. 295–304.
'Reflections on the history of the concept of time'. In *The Study of
    Time I* (ed. J. T. Fraser, F. C. Haber, and G. H. Müller). New York:
    Springer-Verlag, 1972, pp. 1–11.
'Reflections on the natural philosophy of time'. In *New York Academy
    of Sciences, Annals*, Conference on Interdisciplinary Perspectives
    of Time, Vol. 138, Art. 2, 1967, 422–32.
'Time'. *International Science and Technology*, **42** (June 1965), 32–7.
'Time and mathematics'. In *Akten des XVI Internationalen Kongress für
    Philosophie*, Vienna, 2–9 September 1968. Vienna: Herder, pp.
    641–5.

'Time and measurement'. In *Dictionary of the History of Idea: Studies of Selected Pivotal Ideas* (ed. P. P. Wiener). New York: Scribners, 1973, pp. 398–406.

'Time and timing: the astronomical and historical developments'. *Naturwissenschaften*, **54**, 1977, pp. 105–12.

A BIBLIOGRAPHIC SAMPLING CLASSED UNDER THE CHAPTER HEADINGS OF THIS BOOK

1 The origin of our idea of time

Aigner, C., Pochat, G., and Rohsman, A. (ed.). *Zeit/Los: Zur Kunstgeschichte der Zeit*. Cologne: DuMont, 1999.

Argyros, A. J. *A Blessed Rage for Order: Deconstruction, Evolution, and Chaos*. Ann Arbor: University of Michigan Press, 1991.

Barreau, H. *La Construction de la Notion de Temps*. 3 volumes. Strasbourg: Fondemants des Sciences, 1982.

Borst, A. *Computus: Zeit und Zahl im Mittelalter*. Cologne: Bohlau, 1988.

Brann, E. *What, Then, is Time?* Lanham, MD: Rowan and Littlefield, 1999.

Brin, G. *The Concept of Time in the Bible and the Dead Sea Scrolls*. Leiden: Brill, 2001.

Capek, M. (ed.). *The Concepts of Space and Time: Their Structure and their Development*. Dordrecht: Reidel, 1976.

Carr, D. *Time, Narrative, and History*. Bloomington: Indiana University Press, 1986.

Duhem, P. *Medieval Cosmology: Theories of Infinity, Place, Time, Void and the Plurality of Worlds*. University of Chicago Press, 1987.

Dux, G. *Die Zeit in der Geschichte: Ihre Entwicklungslogik von Mythos zur Weltzeit*. Frankfurt am Main: Suhrkamp, 1989.

Ekeland, I. *Mathematics and the Unexpected*. University of Chicago Press, 1988.

Krieger, L. *Time's Reasons: Philosophies of History, Old and New.* University of Chicago Press, 1989.

Levinas, E. *Dieu, La Mort et le Temps.* Paris: Grasset, 1993.

Lippincott, K. (ed.). *The Story of Time.* London: Merrell Holberton, 2000.

Morson, G. S. *Narrative and Freedom: The Shadows of Time.* New Haven: Yale University Press, 1994.

Raulff, U. *Der unsichtbare Augenblick: Zeitkonzepte in der Geschichte.* Göttingen: Wallenstein, 1999.

Romilly, J. de. *Time in Greek Tragedy.* Ithaca: Cornell University Press, 1968.

Sorabji, R. *Time, Creation and the Continuum.* London: Duckworth, 1983.

Ter Meulen, A. G. B. *Representing Time in Natural Language.* Cambridge, MA: MIT Press, 1997.

2 Time and ourselves

Abravaya, I. *Studies of Rhythm and Tempo in the Music of J. S. Bach.* Tel-Aviv University: doctoral thesis, 1999.

Adam, B. *Time and Social Theory.* Philadelphia: Temple University Press, 1990.

Adam, B. *Timewatch: The Social Analysis of Time.* Cambridge: Polity Press, 1995.

Adam, B. *Timescapes of Modernity: The Environment and Invisible Hazards.* London: Routledge, 1998.

Alverson, H. *Semantics and Experience.* Baltimore: Johns Hopkins University Press, 1994.

Assad, M. *Reading with Michel Serres: An Encounter with Time.* Albany: State University of New York Press, 1999.

Bertman, S. *Hyperculture: The Human Cost of Speed.* Westport, CT: Praeger, 1998.

Block, R. A. (ed.). *Cognitive Models of Psychological Time.* Hillside: Erlbaum, 1990.

Cherednichenko, V. I. *Typology of Temporal Relations in Lyrics*. Tbilisi: Academy of Sciences, Georgian SSR, 1986. In Russian.

Coyne, K. *A Day in the Night of America*. New York: Random House, 1992.

Crosby, A. W. *The Measure of Reality: Quantification and Western Society, 1250–1600*. Cambridge University Press, 1997.

Edelman, G. M. *Neural Darwinism: The Theory of Neural Group Selection*. New York: Basic Books, 1987.

Edelman, G. M. *The Remembered Present: A Biological Theory of Consciousness*. New York: Basic Books, 1989.

Edelman, G. M. *Bright Air, Brilliant Fire: On the Matter of the Mind*. New York: Basic Books, 1992.

Emery, E. *Temps et Musique*. Lausanne: L'Age d'Homme, 1998.

Epstein, D. *Shaping Time: Music, the Brain, and Performance*. New York: Schirmer Books, 1995.

Flaherty, M. G. *A Watched Pot: How We Experience Time*. New York University Press, 1999.

Friedman, W. J. *The Developmental Psychology of Time*. New York: Academic Press, 1982.

Geissler, K. A. *Zeit: 'Verweile doch, du bist so schön!'*. Berlin: Quadriga Verlag, 1996.

Geissler, K. A. *Zeit leben. Vom Hasten und Rasten, Arbeiten und Lernen, Leben und Sterben*, 6th edn. Berlin: Quadriga Verlag, 1997.

Gendolla, P. *Zeit: zur Geschichte der Zeiterfahrung*. Cologne: Dumont, 1992.

Gleick, J. *Faster: The Acceleration of Just about Everything*. New York: Pantheon Books, 1999.

Gooddy, W. *Time and the Nervous System*. New York: Praeger, 1988.

Green, A. *Le Temps Eclaté*. Paris: Les Editions de Minuit, 2000.

Grossin, W. *Pour une Science des Temps: Introduction à l'ècologie temporelle*. Toulouse: Octares, 1966.

Heise, U. *Chronoschism: Time, Narrative and Postmodernism*. Cambridge University Press, 1997.

Jackson, J. B. *A Sense of Place, a Sense of Time*. New Haven: Yale University Press, 1994.

Jacobs, M. T. *Short-Term America: The Causes and Cures of our Business Myopia*. Boston: Harvard Business School Press, 1991.

Kaempfer, Wolfgang. *Zeit der Menschen*. Frankfurt: Insel Verlag, 1994.

Kafka, J. S. *Multiple Realities in Clinical Practice*. New Haven and London: Yale University Press, 1989.

Kilwardby, R. *On time and imagination. De Tempore. De Spiritu Fantastico* (ed. Osmund Lewry). Oxford University Press, 1987.

Kramer, J. D. *The Time of Music: New Meanings, New Temporalities, New Listening Strategies*. New York: Schirmer Books, 1988.

León-Portilla, M. *Time and Reality in the Thought of the Maya*. Boston: Beacon Press, 1968.

Macar, F., Pouthas, V., and Friedman, W. J. *Time, Action and Cognition: Towards Bridging the Gap*. Dordrecht: Kluwer Academic, 1992.

Macey, S. L. *The Dynamics of Progress: Time, Method, and Measure*. Athens, GA: University of Georgia Press, 1989.

McFadden, S. H. and Atchley, R. C. (ed.). *Aging and the Meaning of Time*. New York: Springer, 2001.

Melbin, M. *Night as Frontier: Colonizing the World after Dark*. New York: Free Press, 1987.

Michon, J. A., Pouthas, V., and Jackson, J. L. (ed.). *Guyau and the Idea of Time*. Amsterdam: North Holland, 1988.

Nowotny, H. *Time: The Modern and Postmodern Experience*. Oxford: Polity Press, 1994.

Orlock, C. *Inner Time*. New York: Birch Lane Press, 1993.

Pogorilowski, A. *Energies of Musical Time: Essential Studies of Pulsatory Functionalism*. Bucharest: Ararat, 1994.

Rappaport, H. *Marking Time: How our Personalities, our Problems and their Treatments are Shaped by our Anxiety about Time*. New York: Simon & Schuster, 1990.

Read, K. A. *Time and Sacrifice in the Aztec Cosmos*. Bloomington: Indiana University Press, 1998.

Reale, P. *Tempo e Personalità: una Tecnica Psichodiagnostica*. Rome: Bulzoni, 1992.

Reheis, F. *Die Kreativität der Langsamkeit: Neuer Wohlstand durch*

Entschleunigung. Darmstadt: Wissenschaftliche Buchgesellschaft, 1996.

Reiner, T. *Semiotics of Musical Time*. New York: Peter Lang, 2000.

Ricoeur, P. *Time and Narrative*. University of Chicago Press, 1988.

Rizzi, P. *I Percorsi del Tempo: Sulla Psicogenesi della Tempolaità*. Milan: Unicoppli, 1988.

Robinson, J. P. and Godbey, G. *Time for Life: The Surprising Ways Americans Spend their Time*. University Park: Pennsylvania State University Press, 1997.

Schiffer, I. *The Trauma of Time*. New York: International Universities Press, 1978.

Scott, C. E. *The Time of Memory*. Albany: State University of New York Press, 1999.

Shacter, D. L. *Searching for Memory: The Brain, the Mind, and the Past*. New York: Basic Books, 1996.

Sherover, C. *The Human Experience of Time: The Development of its Philosophic Meaning*. New York University Press, 1975.

Slife, B. D. *Time and Psychological Explanation*. Albany: State University of New York Press, 1993.

Stone, R. M. *Dried Millet Breaking: Time, Words, and Song in the Woi Epic of the Kpelle*. Bloomington: Indiana University Press, 1988.

Tarkowska, E. *Time in Polish Life: Results of Research, Hypotheses, Impressions*. Warsaw: Polish Academy of Sciences, 1992. In Polish.

Tedlock, B. *Time and Highland Maya*. Albuquerque: University of New Mexico Press, 1987.

Turner, F. *Shakespeare and the Nature of Time*. Oxford: Clarendon Press, 1971.

Turner, F. *Rebirth of Value: Meditations on Beauty, Ecology, Religion and Education*. Albany: State University of New York Press, 1991.

Weiner, J. *Time, Love, Memory*. New York: Knopf, 1999.

Wendorff, R. (ed.). *Im Netz der Zeit: menschliches Zeiterleben*. Stuttgart: Hirzel, 1989.

Wendorff, R. *Der Mensch und die Zeit*. Opladen: Westdeutscher Verlag, 1988.

Young, M. and Schuller, T. (ed.). *The Rhythms of Society*. London: Routledge, 1988.

Zerubavel, E. *Hidden Rhythms: Schedules and Calendars in Social Life*. University of Chicago Press, 1981.

3 Biological clocks

Edmunds, L. N. *Cellular and Molecular Bases of Biological Clocks*. New York: Springer-Verlag, 1988.

Hildebrandt, G., Moser, M., and Lehofer, M. *Chronobiologie und Chronomedizin: biologische Rhythmen, medizinische Konsequenzen*. Stuttgart: Hippokrates, 1998.

Hughes, M. (ed.). *Body Clock: The Effects of Time on Human Health*. New York: Facts on File, 1989.

Macar, F. *Le Temps: Perspectives Psychophysiologiques*. Brussels: Mardaga, 1980.

Mletzko, I. and Mletzko, H.-G. *Biorhythmik: Elementareinführung in die Chronobiologie*. Wittenberg: Siemens Verlag, 1985.

Mletzko, I. and Mletzko, H.-G. *Die Uhr des Lebens*. Leipzig: Urania Verlag, 1985.

Mletzko, I. and Mletzko, H.-G. *Die Zeit und der Mensch*. Leipzig: Urania Verlag, 1991.

Mletzko, I. and Mletzko, H.-G. *Mensch und Zeit*. Bad Hersfeld: Neuromedizin, 2002.

Palmer, J. D. *Biological Clocks in Marine Organisms*. New York: Wiley-Interscience, 1974.

Palmer, J. D. *The Biological Rhythms and Clocks of Intertidal Animals*. New York: Oxford University Press, 1995.

Palmer, J. D. *The Living Clock: The Orchestrator of Biological Rhythms*. Oxford University Press, 2002.

Rosenberg, G. D. and Runcorn, S. K. (ed.). *Growth Rhythms and the History of the Earth's Rotation*. New York: Wiley, 1975.

Sheving, L. E., Halberg, F., and Pauly, J. E. *Chronobiology*. Georg Thieme: Stuttgart, 1974.

Simakov, K. V. *Origin, Development and Perspectives of the Theory of Paleobiospheric Time*. Magadan: North-East Science Press, 2001.

Smolensky, M. and Lamberg, L. *The Body Clock Guide to Better Health*. New York: Henry Holt, 2000.

Touito, Y. and Haus, E. (ed.). *Biological Rhythms in Clinical and Laboratory Medicine*. New York: Springer-Verlag, 1992.

Wendorff, R. *Die Zeit mit der wir leben*. Jena: Heitkamp, 1991.

Young, M. *The Metronomic Society: Natural Rhythms and Human Timetables*. Cambridge, MA: Harvard University Press, 1988.

4 The measurement of time

Andrewes, W. J. H. *The Quest for Longitude*. Cambridge, MA: Collection of Historical Scientific Instruments, 1996.

Aveni, A. F. *Empires of Time: Calendars, Clocks, and Cultures*. New York: Basic Books, 1989.

Barnett, J. E. *Time's Pendulum: The Quest to Capture Time – From Sundials to Atomic Clocks*. New York: Plenum Press, 1998.

Barreau, H. *Le Temps*. Paris: Presses Universitaires de France, 1996.

Bartky, I. R. *Selling the True Time: Nineteenth Century Timekeeping in America*. Stanford University Press, 2000.

Bear, M. *Days, Months & Years: A Perpetual Calendar*. Norfolk: Tarquin, 1989.

Bedini, S. A. *The Trail of Time: Time Measurement with Incense in East Asia*. New York: Cambridge University Press, 1994.

Blaise, C. *Time Lord: Sir Sandford Fleming and the Creation of Standard Time*. New York, Pantheon, 2000.

Cardinal, C. (ed.). *La Révolution dans la Mesure du Temps: Calendrier Républicain Heures Decimale 1793–1805*. La Chaux-de-Fonds: Musée International d'horlogerie, 1989.

Corbin, A. *Village Bells: Sound and Meaning in the 19th Century French Countryside*. Trans. Martin Thom. New York: Columbia University Press, 1998.

Coyne, G. V., Hoskin, M. A., and Pedersen, O. (ed.). *Gregorian Reform of the Calendar*. Specola Vaticana: Pontificia Academia Scientiarium, 1983.

Doggett, R. (ed.). *Time: The Greatest Innovator. Timekeeping and Time Consciousness in Early Modern Europe*. Washington: Folger Shakespeare Library, 1986.

Duncan, D. E. *Calendar: Humanity's Epic Struggle to Determine a True and Accurate Year*. New York: Avon, 1998.

Gouk, P. *The Ivory Sundials of Nurenberg, 1500–1700*. Cambridge: Whipple Museum of the History of Science, 1988.

Gould, S. J. *Time's Arrow, Time's Cycle*. Cambridge, MA: Harvard University Press, 1988.

Heilbron, J. L. *The Sun in the Church: Cathedrals as Solar Observatories*. Cambridge, MA: Harvard University Press, 1999.

Herberger, C. F. *The Thread of Ariadne: The Labyrinth of the Calendar of Minos*. New York: Philosophical Library, 1972.

King, H. C. *Geared to the Stars: The Evolution of Planetariums, Orreries, and Astronomical Clocks*. University of Toronto Press, 1978.

Landes, D. S. *Revolution in Time: Clocks and the Making of the Modern World*. Cambridge, MA: Harvard University Press, 1983.

Macey, S. L. *Clocks and the Cosmos: Time in Western Life and Thought*. Hamden, CT: Archon Books, 1980.

Macey, S. L. *Patriarchs of Time: Dualism in Saturn-Cronus, Father Time, the Watchmaker God and Father Christmas*. Athens, GA: University of Georgia Press, 1987.

McCrossen, A. *Holiday, Holy Day: The American Sunday*. Ithaca, NY: Cornell University Press, 2000.

Needham, J., Ling, W., and deSolla Price, D. J. *The Hall of Heavenly Records: Korean Astronomical Instruments and Clocks 1380–1780*. Cambridge University Press, 1986.

Needham, J., Ling, W., and deSolla Price, D. J. *Heavenly Clockwork: The Great Astronomical Clocks of Medieval China*. Cambridge University Press, 1986.

Paris, F. (ed.). *The Book of Calendars*. New York: Facts on File, 1982.

Sherman, S. *Telling Time: Clocks, Diaries, and the English Diurnal Form, 1660–1785*. University of Chicago Press, 1996.

Smith, M. M. *Mastered by the Clock: Time, Slavery, and Freedom in the American South*. Chapel Hill: University of North Carolina Press, 1997.

Sobel, D. *Longitude*. New York: Walker, 1985.

Turner, A. J. *Ritmi de Cielo e Misura del Tempo*. Brugine: Edizione 1 + 1, 1985.

Turriano, J. *Breve discurso a su Majestad el Rey Catolico en torno a la reduccion del ano y reforma del calendario*. Madrid: Fundacion Juanelo Turriano, 1990.

Waugh, A. *Time: From Microseconds to Millennia: A Search for the Right Time*. London: Headline Books, 1999.

Wendorff, R. *Tag und Woche, Monat und Jahr: eine Kulturgeschichte des Kalenders*. Opladen: Westdeutscher Verlag, 1993.

Wilcox, D. J. *The Measure of Time Past: Pre-Newtonian Chronologies and the Rhetoric of Relative Time*. University of Chicago Press, 1987.

Zerubavel, E. *The Seven Day Circle: the History and Meaning of the Week*. New York: Free Press, 1985.

5 Time and relativity

Bernard, P. *Philosophie et Science du Temps*. Paris: PUF, 1996.

Friedman, W. J. *About Time: Inventing the Fourth Dimension*. Cambridge, MA: MIT Press, 1990.

Gott III, J. R. *Time Travel in Einstein's Universe: The Physical Possibilities of Travel through Time*. New York: Houghton Mifflin, 2001.

Kazaryan, V. P. *The Concept of Time in Philosophy and Physics*. Moscow: Nauka, 1980. In Russian.

Marder, L. *Time and the Space-Traveller*. London: Allen & Unwin, 1971.

Nahin, P. J. *Time Machines: Time travel in Physics, Metaphysics and Science Fiction*. New York: American Institute of Physics, 1993.

6 Time, gravitation and the universe

Barrow, J. D. *Between Inner Space and Outer Space*. New York: Oxford University Press, 1999.

Davies, P. *About Time*. New York: Simon & Schuster, 1995.

Fagg, L. W. *Two Faces of Time*. Wheaton: Quest, 1985.

Gorst, M. *Measuring Eternity: The Search for the Beginning of Time*. New York: Broadway, 2001.

Hawking, S. W. *A Brief History of Time*, updated and expanded 10th anniversary edn. New York: Bantam, 1998.

Layzer, D. *Cosmogenesis: The Growth of Order in the Universe*. New York: Oxford University Press, 1990.

Mittelstaedt, R. *Der Zeitbegriff in der Physik*. Mannheim: Bibliographische Institut, 1980.

Park, D. *The Image of Eternity: Roots of Time in the Physical World*. Amherst: University of Massachusetts Press, 1980.

Thorn, K. S. *Black Holes and Time Warps: Einstein's Outrageous Legacy*. New York: W. W. Norton, 1994.

7 The origin and arrow of time

Fagg, L. W. *The Becoming of Time*. Athens, GA: Scholars Press, 1995.

Halliwell, J., Pérez-Mercader, J., and Zurek, W. H. (ed.). *Physical Origins of Time Asymmetry*. Cambridge University Press, 1994.

Horwich, P. *Asymmetries in Time*. Cambridge, MA: MIT Press, 1987.

Lebedev, Y. A. *The Ambivalent Universe: Apocryphal Reflections upon Time's Arrows*. Moscow: Kostroma, 2000. In Russian.

Lestienne, R. *Le Hasard Créateur*. Paris: Éditions la Découverte, 1993.

Lestienne, R. *The Children of Time: Causality, Entropy, Becoming*. Urbana: University of Illinois Press, 1995.

Price, H. *Time's Arrow and Archimedes' Point*. Oxford University Press, 1996.

Savitt, S. (ed.). *Time's Arrows Today*. Cambridge University Press, 1995.

8 The significance of time

Dolnikowski, E. W. *Thomas Bradwardine: A View of Time and a Vision of Eternity in Fourteenth-Century Thought.* Leiden: Brill, 1995.

Geissler, K. A. *Vom Tempo der Welt: am Ende der Uhrzeit.* Freiburg: Herder, 1999.

Giulio, L. F. *Saggio di una Cronologia delle Idee e delle Esperienze sul Tempo.* Bologna: CLUEB, 2000.

Halpern, P. *Time Journeys: A Search for Cosmic Destiny and Meaning.* New York: McGraw-Hill, 1990.

Helm, B. P. *Time and Reality in American Philosophy.* Amherst: University of Massachusetts Press, 1985.

Hörz, H. *Philosophie der Zeit.* Berlin: Verlag der Wissenschaften, 1989.

Molchanov, Y. B. *Four Conceptions of Time in Philosophy and Physics.* Moscow: Nauka, 1977. In Russian.

Rinderspracher, J. P. *Gesellschaft ohne Zeit.* Frankfurt: Campus Verlag, 1985.

Rosenthal, S. B. *Time, Continuity, and Indeterminacy: A Pragmatic Engagement with Contemporary Perspectives.* Albany: State University of New York Press, 2000.

Sandbothe, M. *Die Verzeitlichung der Zeit: Grundtendenzen der modernen Zeitdebatte in Philosophie und Wissenschaft.* Darmstadt: Wissenschaflitche Buchgesellschaft, 1998.

Sherover, C. M. *Time, Freedom, and the Common Good: An Essay in Public Philosophy.* Albany: State University of New York Press, 1990.

EDITED VOLUMES

Achtner, W., Kunz, S., and Walter, T. (ed.). *Dimensionen der Zeit: Die Zeitstrukturen Gottes, der Welt und des Menschen.* Darmstadt: Primus, 1998.

Aveni, A. F. (ed.). *World Archeoastronomy: Selected Papers from the 2nd*

Oxford International Conference on Archeoastronomy. Cambridge University Press, 1989.

Baudson, M. (ed.). *Zeit: Die Vierte Dimension in der Kunst*. Weinheim: Acta Humaniora, 1985.

Biervert, B. and Held, M. (ed.). *Zeit in der Ökonomik: Perspektiven für die Theoriebildung*. Frankfurt: Campus Verlag, 1995.

Buccheri, R., di Gesu, V., and Saniga, M. (ed.). *Studies on the Structure of Time: From Physics to Psychopathology*. New York: Kluwer Academic/Plenum, 2000.

Buccheri, R. and Saniga, M. (ed.). *The Nature of Time: Geometry, Physics and Perception*. Moscow: Kluwer Academic/Plenum, 2002.

Butterfield, J. (ed.). *The Arguments of Time*. Oxford University Press, 1999.

Cardinal, C., Jequier, F., Barrelet, J.-M., and Beiner, A. (ed.). *L'Homme et le Temps en Suisse, 1291–1991*. La Chaux-de-Fonds: Musée International d'horlogerie, 1991.

Castelli, E. (ed.). *Il Tempo*. Padova: CEDAM, 1958.

Cleary, T. R. (ed.). *Time, Literature and the Arts: Essays in Honor of Samuel L. Macey*. Victoria, BC: University of Victoria Press, 1994.

Fechtrup, H., Schulze, F., and Sternberg, T. (ed.). *Zwischen Anfang und Ende: Nachdenken über Zeit, Hoffnung und Geschichte*. Münster: Lit Verlag, 2000.

Forman, F. J. and Sowton, C. (ed.). *Taking our Time: Feminist Perspectives on Temporality*. Oxford: Pergamon Press, 1989.

Fraser, J. T. (ed.). *The Voices of Time: A Cooperative Survey of Man's Views of Time as Expressed by the Sciences and the Humanities*. New York: Braziller, 1966, and London: Allen Lane the Penguin Press, 1968; 2nd edn, Amherst: University of Massachusetts Press, 1981. [An ebook at <http://www.netLibrary.com>.]

Fraser, J. T., Haber, F. C., and Müller, G. H. (ed.). *The Study of Time I*. New York: Springer-Verlag, 1972.

Fraser, J. T. and Lawrence, N. (ed.). *The Study of Time II*. New York: Springer-Verlag, 1975.

Fraser, J. T., Lawrence, N., and Park, D. (ed.). *The Study of Time III*. New York: Springer-Verlag, 1978.

Fraser, J. T., Lawrence, N., and Park, D. (ed.). *The Study of Time IV*. New York: Springer-Verlag, 1981.

Fraser, J. T. and Lawrence, N. (ed.). *Time. Science, and Society in China and the West: The Study of Time V*. Amherst: University of Massachusetts Press, 1986. [An ebook at <http://www. netLibrary.com>.]

Fraser, J. T. (ed.). *Time and Mind: The Study of Time VI*. Madison, CT: International Universities Press, 1989.

Fraser, J. T. and Rowell, L. (ed.). *Time and Process: The Study of Time VII*. Madison, CT: International Universities Press, 1992.

Fraser, J. T. and Soulsby, M. P. (ed.). *Dimensions of Time and Life: The Study of Time VIII*. Madison, CT: International Universities Press, 1995.

Fraser, J. T., Soulsby, M. P., and Argyros, A. (ed.). *Time, Order, Chaos: The Study of Time IX*. Madison, CT: International Universities Press, 1998.

Fraser, J. T. and Soulsby, M. P. (ed.). *Time at the Millennium: Changes and Continuities: The Study of Time X*. Westport, CT: Bergen & Garvey, 2000.

Gimmler, A., Sandbothe, M., and Zimmerli, W. C. (ed.). *Die Wiederentdeckung der Zeit: Reflexionen, Analysen, Konzepte*. Darmstadt: Primus, 1997.

Heinemann, G. (ed.). *Zeitbegriffe: Zeitbegriff der Naturwissenschaften, Zeiterfahrung und Zeitbewusstsein*. Freiburg: Karl Alber, 1986.

Heintel, P. and Macho, T. (ed.). *Zeit und Arbeit hundert Jahre nach Marx*. Wien: Wissenschaftliche Gesellschaft Osterreichs. 1985.

Held, M. and Geissler, K. A. (ed.). *Ökologie der Zeit: vom Finden der Rechten Zeitmasse*. Stuttgart: Hirzel, 1993.

Held, M. and Geissler, K. A. (ed.). *Von Rhythmen und Eigenzeiten: Perspektiven einer Ökologie der Zeit*. Stuttgart: Hirzel, 1995.

Hersch, J. and Poirier, R. (ed.). *Entretiens sur le Temps*. Paris: Mouton, 1967.

Hughes, D. O. and Trautman, T. (ed.). *Time: Histories and Ethnologies*. Ann Arbor: University of Michigan Press, 1995.

Levich, A. P. (ed.). *On the Way to Understanding the Time Phenomenon:*

The Construction of Time in Natural Science. 2 volumes. Singapore: World Scientific, 1995–96.

Mallmann, C. A. and Nudler, O. (ed.). *Time, Culture, and Development.* Bariloche, Argentina: Fundacion Bariloche, 1986.

Oaklander, L. N. (ed.). *The Importance of Time: Proceedings of the Philosophy of Time Society, 1995–2000.* Dordrecht: Kluwer, 2001.

Paflik, H. (ed.). *Das Phänomen der Zeit in Kunst und Wissenschaft.* Weinheim: Acta Humaniora, 1987.

Paisl, A. and Mohler, A. (ed.). *Die Zeit.* Munich: Oldenburg, 1983.

Pfusterschmid-Hardtenstein, H. (ed.). *Zeit und Wahrheit.* Vienna: Ibera, 1995.

Rinderspacher, J. P. (ed.). *Zeit für die Umwelt: Handlungskonzepte für eine ökologische Zeitverwendung.* Berlin: Sigma, 1996.

Sabbadini, A. (ed.). *Il Tempo in Psicoanalisi.* Milan: Feltrinelli, 1979.

Schall, J. (ed.). *Tempus Fugit: An Exhibition Catalog.* Kansas City, MO: Nelson–Atkins Museum of Art, 2000.

Szalai, A. and Converse, P. E. (ed.). *The Use of Time.* The Hague: Mouton, 1972.

Tholen, G. C., Scholl, M., and Heller, M. (ed.). *Zeitreise; Bilder/Maschinen/Strategien/Rätsel.* Basel: Stroemfeld/Roten Stern, 1993.

Turner, A. J. (ed.). *Time: An Exhibition Catalog.* The Hague: Tijd voor Tijd Foundation, 1990.

Von Auer, F., Geissler, K., and Schauer, H. (ed.). *Auf der Suche nach gewonnener Zeit.* Mössingen-Talheim: Talheimer Verlag, 1990.

Weis, K. (ed.). *Was treibt die Zeit: Entwicklung und Herrschaft der Zeit in Wissenschaft, Technik und Religion.* Munich: DTV, 1998.

Wieck, R. S. (ed.). *Time Sanctified: The Book of Hours in Medieval Art and Life.* New York: Braziller, 1988.

Zajaczkowski, A. (ed.). *Czas w Kulturze.* Warsaw: Panstwowy Insytut Wydawniczy, 1988.

Zoll, R. (ed.). *Zerstörung und Wiederaneignung der Zeit.* Frankfurt: Suhrkamp, 1988.

THE INTEGRATED STUDY OF TIME: BOOKS BY J. T. FRASER

Time, Conflict, and Human Values. Urbana and Chicago: University of Illinois Press, 1999.

Of Time, Passion, and Knowledge: Reflections on the Strategy of Existence. 2nd edn. Princeton University Press, 1990.

Time the Familiar Stranger. Amherst: University of Massachusetts Press, 1987. Also in German, Italian, and English Braille.

The Genesis and Evolution of Time: A Critique of Interpretations in Physics. Amherst: University of Massachusetts Press, 1982. Also in Japanese and Spanish.

Time as Conflict: A Scientific and Humanistic Study. Basel and Boston: Birkhäuser, 1978.

INDEX